Active Control of
Aircraft Cabin Noise

Computational and Experimental Methods in Structures

Series Editor: Ferri M. H. Aliabadi *(Imperial College London, UK)*

Vol. 1 Buckling and Postbuckling Structures: Experimental, Analytical and Numerical Studies
edited by B. G. Falzon and M. H. Aliabadi (Imperial College London, UK)

Vol. 2 Advances in Multiphysics Simulation and Experimental Testing of MEMS
*edited by A. Frangi, C. Cercignani (Politecnico di Milano, Italy),
S. Mukherjee (Cornell University, USA) and N. Aluru (University of Illinois at Urbana Champaign, USA)*

Vol. 3 Multiscale Modeling in Solid Mechanics: Computational Approaches
edited by U. Galvanetto and M. H. Aliabadi (Imperial College London, UK)

Vol. 4 Boundary Element Methods in Engineering and Sciences
*by M. H. Aliabadi (Imperial College, UK) and
P. Wen (Queen Mary University of London, UK)*

Vol. 5 Mathematical Methods and Models in Composites
edited by V. Mantic (University of Seville, Spain)

Vol. 6 Woven Composites
edited by M. H. Aliabadi (Imperial College London, UK)

Vol. 7 Active Control of Aircraft Cabin Noise
*by I. Dimino (CIRA, The Italian Aerospace Research Centre, Italy)
and M. H. Aliabadi (Imperial College London, UK)*

Computational and Experimental Methods in Structures – Vol. 7

Active Control of Aircraft Cabin Noise

Ignazio Dimino
CIRA, The Italian Aerospace Research Centre, Italy

Ferri Aliabadi
Imperial College London, UK

Imperial College Press

Published by

Imperial College Press
57 Shelton Street
Covent Garden
London WC2H 9HE

Distributed by

World Scientific Publishing Co. Pte. Ltd.
5 Toh Tuck Link, Singapore 596224
USA office: 27 Warren Street, Suite 401-402, Hackensack, NJ 07601
UK office: 57 Shelton Street, Covent Garden, London WC2H 9HE

Library of Congress Cataloging-in-Publication Data
Dimino, Ignazio.
    Active control of aircraft cabin noise / Ignazio Dimino, CIRA, The Italian Aerospace Research Centre, Italy, Ferri Aliabadi, Imperial College London, UK.
    pages cm. -- (Computational and experimental methods in structures ; volume 7)
    Includes bibliographical references and index.
    ISBN 978-1-78326-657-9 (hardcover : alk. paper)
    1. Aircraft cabins--Noise. 2. Acoustical engineering. I. Aliabadi, M. H. II. Title.
    TL671.65.D56 2015
    629.134'42--dc23
                                          2014050210

British Library Cataloguing-in-Publication Data
A catalogue record for this book is available from the British Library.

Copyright © 2015 by Imperial College Press

All rights reserved. This book, or parts thereof, may not be reproduced in any form or by any means, electronic or mechanical, including photocopying, recording or any information storage and retrieval system now known or to be invented, without written permission from the Publisher.

For photocopying of material in this volume, please pay a copying fee through the Copyright Clearance Center, Inc., 222 Rosewood Drive, Danvers, MA 01923, USA. In this case permission to photocopy is not required from the publisher.

Typeset by Stallion Press
Email: enquiries@stallionpress.com

Printed in Singapore

to Major Giuseppe Palminteri

Contents

Preface   xi

Nomenclature   xv

1 Aircraft Noise Control   1
    1.1 Introduction   1
    1.2 Active Noise Control and Active
        Structural Acoustic Control   7
        1.2.1 Active noise control applications in aircraft   9
    1.3 Book Overview   12

2 Fundamentals in Structural Acoustics   17
    2.1 Introduction   17
    2.2 Metrics of Sound Transmission   19
        2.2.1 Sound pressure level   20
        2.2.2 Sound insertion loss   21
        2.2.3 Sound transmission loss   22
    2.3 Transmission Loss Experiments   23
    2.4 Experimental Modal Analysis   25
        2.4.1 Basics in structural dynamics   26
        2.4.2 Modal parameter estimation algorithms   28
    2.5 The Acoustic Wave Equation and the Helmholtz
        Equation   29
        2.5.1 The inhomogeneous wave equation   31
        2.5.2 The solution to the inhomogeneous wave equation   31

		2.5.3	Free space sound radiation	33
		2.5.4	The Kirchoff–Helmholtz integral equation	34
	2.6	Sound Radiation from Vibrating Flat Panels		34
		2.6.1	Sound power radiation	37
		2.6.2	Elemental radiators	39
		2.6.3	Radiation modes	41
		2.6.4	Sound radiation efficiency	42
	2.7	Finite Element Method for Interior Problems		44
	2.8	Coupled FE Formulation for Interior Vibro-Acoustic Systems		49
	2.9	Summary		51
3	**Transmission of Sound through Multiple Partitions**			**53**
	3.1	Introduction		53
	3.2	Analytical Modelling of Multi-Panel Partitions		55
		3.2.1	Infinite panels	55
		3.2.2	Modal coupling theory	59
		3.2.3	Structure-acoustic modal coupling in triple-panel partitions	63
	3.3	Sound Transmission Through Infinite Triple Partitions		69
		3.3.1	Analysis of the derived equations	73
	3.4	Sound Transmission Simulation		79
		3.4.1	Diffuse acoustic excitation model	81
		3.4.2	Benchmark examples	82
	3.5	Summary		91
4	**Adaptive Control of Sound Radiation**			**93**
	4.1	Introduction		93
	4.2	Control of Sound Radiation by Structural Actuators		97
	4.3	Control Strategies		101
		4.3.1	Feedforward control	101
		4.3.2	Feedback control	102
		4.3.3	Feedforward vs. feedback control	104
	4.4	Steepest-Descent Algorithm		105
	4.5	Adaptive Digital Filters		107
	4.6	Filtered-X LMS Algorithm		110
	4.7	Multi-Channel Adaptive Algorithms		111

CONTENTS ix

    4.8 Experimental System Identification 114
    4.9 Numerical Simulations 118
          4.9.1 Harmonic primary noise 122
          4.9.2 Random primary noise 125
          4.9.3 Narrow-band primary noise 130
    4.10 Summary 135

5 Noise-Reducing Smart Windows   139

    5.1 Introduction 139
    5.2 Literature Survey 143
          5.2.1 Control of sound transmission through
                multi-wall partitions: panel control 146
          5.2.2 Control of sound transmission through
                multi-wall partitions: cavity control 148
    5.3 Piezoelectricity 151
    5.4 Smart Window Design 156
          5.4.1 Scope 158
          5.4.2 Numerical modelling 160
          5.4.3 Structural control actuators 165
    5.5 Smart Window Test-Bed 172
          5.5.1 Experimental modal analysis 174
          5.5.2 Control actuators set-up 177
    5.6 Transmission Loss Predictions 186
    5.7 Summary 187

6 Active Noise Control Experiments   189

    6.1 Introduction 189
    6.2 Experimental Set-Up 191
    6.3 Instrumentation 195
    6.4 Imperfections in the Experimental Set-Up 196
          6.4.1 Coupled vibro-acoustic model of the sending
                room 198
          6.4.2 Experimental validation 202
    6.5 Real-Time DSP Implementation of Active Noise
          Control 206
          6.5.1 The real-time controller 207
          6.5.2 Experimental results 211

	6.6	Sound Intensity Measurements	231
		6.6.1 Experimental results	233
	6.7	Passenger Comfort Estimation	239
		6.7.1 Subjective noise perception	239
		6.7.2 Virtual Passenger Model	241
	6.8	Summary	243
A	**Benchmark Examples**		**245**
	A.1	Introduction	245
	A.2	Laminar Piezoelectric Sensor	245
	A.3	Cantilever Piezoelectric Plate	247
		A.3.1 Static analysis	247
		A.3.2 Dynamic analysis	249
B	**Sound Power Simulations**		**253**
	B.1	Radiation Efficiency	253
	B.2	Radiation Efficiency of the Window Test-Bed	255
C	**Preliminary Experimental Studies**		**259**
Bibliography			263
Index			279

Preface

The adaptation of engineered structures to suit the working environment has historical examples dating back to Leonardo da Vinci. However, practical applications of systems capable of adapting to and optimizing themselves for changing operating conditions are relatively recent and still emerging.

Smart structures have enabled higher performance, safer and energy-efficient systems incorporating sensors and actuators in combination with electronics in order to autonomously sense the environment and adaptively respond to an external stimulus under a variety of operational conditions. The advent of cost-effective digital signal processing hardware has been the principal enabler of their practical applications in the aerospace, mechanical and civil engineering disciplines and has driven the intensive interest and excitement in recent years.

Design and control of smart structures is a highly interdisciplinary field encompassing smart materials, actuation and sensing, control strategies and electronics. Cross-disciplinary studies bringing together experts from different sciences have become a focal point of modern research, regularly combining insights from modelling, analysis and experiments.

Although the applications and goals are numerous and their scale and benefits may vary from one application to the other, the use of smart materials, such as piezoelectric ceramics, magnetorheological fluids and shape memory alloys, for which there is already a copious literature, is a common practice in actual and theoretical studies. On the whole, these research topics share a growing potential in noise and vibration control, ranging from low-frequency noise attenutation devices in aircraft engines and road vehicle suspension systems to protect aircraft cabins and car interiors, to adaptive systems for mitigating aircraft noise pollution affecting airports located close to dense population areas.

As vibro-acoustic comfort is increasingly emerging as one of the key market factors in the aviation industry, new design paradigms that slightly deviate from the effort in exclusively reducing aircraft weight are seeing growing interest. To overcome the challenges of providing light-weight structures, motivated by potential fuel savings on the one hand but inducing higher vibro-acoustic levels on the other, the two targets being totally antipodal, concepts based on smart materials and structures are currently being investigated. They provide new design alternatives for interior noise control which may impact on passengers from the point of view of comfort perception. Their novelty relies on considering passenger comfort metrics while designing aircraft interior sound quality and making it adaptive with aircraft flight configuration, speed, weight, etc.

Driven by the need for ever lighter and more acoustically efficient aeronautical structures, active structural control offers a viable solution to adaptively enhance aircraft interior noise. A suitable compromise for controlling sound radiation into aircraft cabins may be achieved by beneficially changing the inherent structural characteristics of the fuselage panels in response to an external disturbance. Sensors and actuators may be integrated into the structure itself and driven by an intelligent system to enhance noise radiation properties by combining digital signal processing techniques with traditional acoustics.

Despite widespread availability of smart solutions capable of enhancing aircraft cabin comfort, active structural control remained at the study phase with little implementation for many years. There were many reasons. Firstly, microprocessors for active control using adaptive digital signal processing had not yet evolved as effective and inexpensive solutions, as they are today. Secondly, there was a lack of suitable actuators for low-frequency active control of heavy structures with reasonable power consumption and low weight and volume. Thirdly, enhancements in passive noise control technology were often less risky and more effective than active control methods. However, although active noise and vibration control is emerging as a competitive solution for enhanced cabin comfort, there are still important technical issues to be solved in order to move from theoretical studies to practical applications in aircraft cabins. Furthermore, the perceived noise field in the cabin may be more accurately assessed if performance metrics related to specific human perception attributes are taken into account from the early design phase of the aircraft by considering both noise and vibration transmission paths.

To highlight some of the recent achievements that have propelled the use of active noise and vibration control systems, this book offers a complete guide and reference to modern interior noise control principles implemented in smart structures to reduce unwanted noise and vibration. Particular emphasis is given to the design, modelling and testing of smart panels for aircraft cabin noise control. In a broader perspective, noise and vibration-related design aspects involving structural acoustics are extensively addressed by demonstrating the potential impact of active control on cabin interior comfort perceived by passengers in aircraft-localized areas. Benefits associated with practical implementations of adaptive concepts are validated by experiments involving sensors, actuators and electronic devices, thus enabling the reader to compare more acoustically efficient active noise control systems with the passive counterparts. Control of sound radiation and transmission with adaptive structures is dealt with along with recommended practices for noise control experiments.

This book is planned as a reference publication, easily readable by graduate students and even ambitious undergraduates in aeronautical engineering. It also serves as an authoritative resource book for research into the subject, although theoretical and experimental research is comprehensive enough to capture the interest of non-experts in engineering who wish to have an overview of some of the active noise control applications in aircraft. The book features focal aims of active and passive noise control of single- and multi-panel systems. It provides recommended methodology to evaluate sound transmission-related problems and suggests approaches to their adaptive solutions without extensive tables and formulations. Theoretical, experimental and applied aspects are covered, including the main formulations and comparisons of theory and experiments.

Nomenclature

Letters

Γ_Z	acoustic boundary surface with a prescribed normal impedance distribution
Γ_p	acoustic boundary surface with a prescribed pressure distribution
Γ_V	acoustic boundary surface with a prescribed normal velocity distribution
Ω_a	acoustic domain
Ξ	acoustic potential energy
Γ_0	acoustically rigid boundary surface
V	acoustic volume
k	acoustic wavenumber
ρ_0	ambient fluid density
A	area of the panel
D	bending stiffness
D_1	bending stiffness of the incident panel
D_2	bending stiffness of the middle panel
D_3	bending stiffness of the radiating panel
Γ	boundary surface
ω	circular frequency
μ	convergence factor
S_a	cross sectional area of the actuator
E_a	elastic modulus of a piezoelectric actuator
D	electric charge per unit area

E	electric field
e	electromechanical coupling
$\mathrm{E}\,[\cdot]$	expectation operator
ρ_F	fluid dynamic density
P	fluid dynamic pressure
v_F	fluid velocity
q	fluid volume velocity per unit volume
d	generic distance between two panels
k_s	host structure stiffness
Π_i	incident sound power
l_m	larger in-plane dimension of the radiating panel
l_a	length of the stack actuator
m_2	mass of the glass panel
m_1	mass of the plexiglass panel
ρ_{s_1}	mass per unit area of the incident panel
ρ_{s_2}	mass per unit area of the middle panel
ρ_{s_3}	mass per unit area of the radiating panel
n	number of disks of a piezoelectric stack actuator
h	partition thickness
ϵ	permittivity
k_p	piezoelectric actuator stiffness
c	speed of sound
S	strain vector
T	stress vector
ρ_s	structural mass per unit volume
Γ_S	structural boundary surface with a prescribed normal displacement distribution
Π_t	transmitted sound power
ϕ	voltage

Acronyms

ANC	active noise control
ANN	artificial neural network
ASAC	active structural acoustic control
DSP	digital signal processing
FRF	frequency response function
LMMS	least maximum mean squares algorithm

MELMS	multi-channel filtered-X LMS
MELMS	multiple error LMS algorithm
MIMO	multi-input-multi-output
MISO	multi-input-single-output
NVH	noise, vibration and harshness
SIMO	single-input-multi-output
SISO	single-input-single-output
STFT	short time fourier transform
TRL	Technology Readiness level

Chapter 1

Aircraft Noise Control

1.1 Introduction

The commercial aviation industry has achieved drastic improvements over the last few decades, thanks to research efforts and technology advances. Novel technologies and enhanced low noise operational procedures have made aircraft quieter and reduced the population affected by noise, and that is forecast to continue.

Aircraft noise levels have benefited from continuous enhancements in technologies enabling engine noise reduction at source. Advances in materials and manufacturing processes have allowed modern turbofans to deliver significant reductions in aircraft noise, while simultaneously reducing fuel burn [1, 2]. With the development of quieter engines characterized by high-bypass ratio and acoustic liners in the nacelle, airframe noise has become a major noise source, and its impact on interior noise remains a challenging topic. The progress already made in aircraft noise emission is shown in Fig. 1.1 [3].

Further design technologies are on the horizon to deliver minimum airframe noise [4–6]. In this regard, novel attenuation methods currently being developed using perforated metals and acoustic liners provide perspectives on potential opportunities for applications on high-lift devices. Passive porous treatments as well as flap side-edge fences have demonstrated to be effective in noise suppression by causing a reduction in the strength of the flap tip vortex [7, 8]. Slat noise attentuation using absorptive acoustic liners has also shown encouraging results. The physical mechanism of flap side-edge noise generation has been widely investigated and it is associated

2   CHAPTER 1. AIRCRAFT NOISE CONTROL

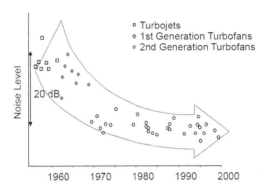

Figure 1.1. Progress in the field of aircraft noise emission over the last few decades [3]

with the pressure jump across the upper and lower surfaces of the flap that creates a re-circulating flow around the side-edge. A lined flap with micro-perforated sheets and internal honeycomb layers does not incur any specific aerodynamic penalties and could easily be retrofitted to existing airplanes. Similarly to Helmholtz resonators, the cavities generated by micro-holes and honeycomb allow the attenuation of narrow-band noise generated at the flap trailing edge, especially during approach to the landing phase of the aircraft operation [9]. Alternatively, adaptive trailing edge concepts enabling wing camber variations in response to changing flying speeds can help make aircraft safer, quieter and more efficient [10, 11]. The key technological advantage is that the wing is deformed to conform to the noise-optimized position in different flight configurations by eliminating gaps in wing edges which generate noise and makes flight less efficient.

Although aircraft are nowadays more than 20 dB quieter than in the past, these benefits are not fully perceived by aircraft passengers. Passenger demands for enhanced cabin comfort, along with the tightening of legislation on noise pollution and human exposure to noise, have made aircraft interior noise an important commercial asset and one of the primary market catalysts. While the regulation of noise around airports is becoming stricter and stricter, the competition to provide more comfortable aircraft has encouraged manufacturers to invest in the use of smart structures to reduce unwanted noise and vibration. However, unlike the great strides made in reducing noise impact on communities exposed to noise from air transport activites, design and certification criteria on aircraft interior noise and vibration are still contradictory and in the process of stabilizing. Comfort assessment in accordance with ISO 2631 [12] dealing with human

1.1. INTRODUCTION

exposure to vibration is often ineffective in evaluating passenger comfort perception associated with aircraft sources. Revised operating noise limitations would require additional technological advances in airframe, wing and engine design [13].

It is well established that the most salient sources of fixed-wing aircraft interior noise are the propulsion system, being either a propeller or turbofan engine, the turbulent boundary layer (TBL) and the aircraft structure itself [14, 15]. Airframe noise is mainly generated by flap side-edges, slats and landing gear. Recent works indicate that the landing gear with high-lift devices contributes to approximately 30% of the overall noise emission of the aircraft during take-off and landing phases [16]. The flow separation over the landing gear components, such as hydraulic cables, electrical wiring, torque links and front and rear braces, which for ease of inspection and maintenance are exposed to air flow, constitutes the primary broadband noise signature covering frequencies from approximately 90 Hz to 4 kHz.

Although the individual contributions to the overall level of noise may vary among aircraft components, interior noise levels are often above the threshold of 80 dB at which permanent hearing damage can potentially occur. Differences may also arise between aircraft with wing-mounted and fuselage-mounted engines. Both turbulence in the boundary layer on the outer surface of the cabin and propulsion-system-induced engine sound are transmitted into the aircraft via the air (airborne), Fig. 1.2, and via

Figure 1.2. Sound pressure level distribution due to monopole sources simulating fan noise of a Dassault Falcon [19]

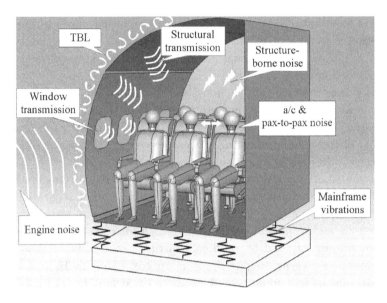

Figure 1.3. Major sources of noise and vibration in a human-centred cabin design approach

the airframe structure (structure-borne), Fig. 1.3. Airborne-noise radiates directly from the source. Jet noise typically peaks in the direction of the jet flow and it is a significant broadband contributor to middle- and high-frequency sound pressure levels in the cabin. Landing gear noise spectrum can be roughly decomposed into three frequency components, namely the low-, mid- and high-frequency components, representing contributions from the wheels, the main struts and the small dresses, respectively. As an example, the individual noise sources of a fully dressed landing gear model are shown in Fig. 1.4 [17]. Structure-borne noise is radiated from the fuselage panels due to the random excitation caused by air turbulence and due to the tonal frequencies caused by the rotating imbalances within the engines. Both the jet and propeller engines produce discrete tones and broadband noise. Fuselage panels radiate sound because of external acoustic excitation, causing the surfaces to vibrate, and absorb energy from the incident noise field because of the finite surface acoustic impedance. The distinction between the incident field and the radiated field is not obvious, even in the near field, and recently, novel measurement methods based on microphone arrays, such as acoustic holography or beamforming, have been proposed to experimentally identify noise sources and transmission paths [18].

1.1. INTRODUCTION

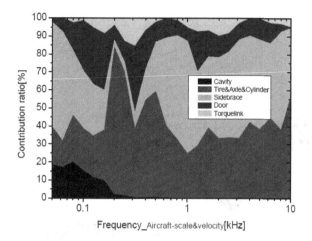

Figure 1.4. Noise sources of a two-wheel landing gear model [17]

Though jets are expected to dominate the market during the next ten years and beyond, global trends in rising fuel prices are leading to a renewed interest in open rotor and turboprop aeroengines for the regional aviation market. Turboprop engines have lower fuel consumption per passenger and CO_2 emissions than comparable turbofan engines but higher engine maintenance costs, shorter ranges and lower cruising speeds. Moreover, despite their fuel efficiency, the radiated noise is more significant for open rotor engines than conventional turbofans since the rotors are fully exposed to oncoming turbulence and lack ducting to attenuate the radiated sound.

In turbopropeller-driven aircraft, cabin noise has long been recognized as a potential cause of passenger discomfort. Interior noise spectra are dominated by low-frequency discrete tones occuring at integral multiples of the blade passage frequency (BPF), which is generally in the order of 100 Hz. Figure 1.5 shows an exemplary interior noise spectrum up to 600 Hz for a typical single aisle aircraft. Noise levels, often in excess of 80 dB, are typically 5–10 dB higher than those in a comparable turbofan-powered aircraft. Since low-frequency sound is more damaging and usually perceived as very annoying by passengers, most of the research applied to interior noise in turboprop aircraft has been devoted towards the control of low-frequency narrow-band excitations.

Several methods for controlling airborne and structure-borne aircraft noise are currently being pursued. They range from conventional fuselage soundproofing methods to more sophisticated techniques such as

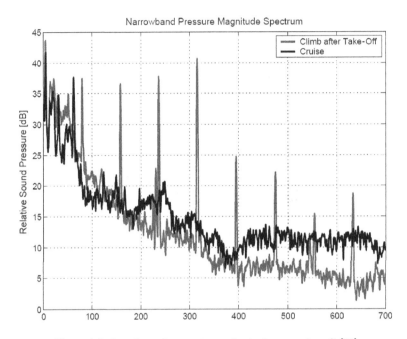

Figure 1.5. Interior noise spectrum of a turboprop aircraft [20]

synchrophased propellers in turboprop aircraft. Some of these technologies are at higher technology readiness level (TRL) than others. Fairings or perforated fairings, for instance, are mature technologies for existing landing gear retrofits. Others are relatively new and require further refinements to be integrated into real structures.

Sound absorbing materials are effective in reducing vibration and noise transmission through the fuselage at mid and high frequencies. At low frequencies, they feature an unacceptable increase in mass and volume, as their thickness shall be comparable or at least one quarter of the sound wavelength. Tuned vibration absorbers (TVAs) have been successfully used for some time to control structure-borne noise associated with engine vibrations. The Fokker F27, for instance, used TVAs attached directly to the fuselage frames and tuned to low BPF harmonics, in order to attenuate cabin interior noise levels [21]. As a drawback to such a benefit, they worked well only at a single frequency. Nevertheless, due to the aircraft's varying operating conditions, the absorber's frequency and damping require adaptability to allow for varying engine speed and to handle several frequencies simultaneously.

1.2. ACTIVE NOISE CONTROL

The real breakthrough in the research on active control technologies began in the 1980s with the progress in active and semi-active control through digital signal processing (DSP) techniques along with a cost saving design of smart material-based systems. Smart systems utilizing piezoelectric materials to suppress structural vibrations and control noise transmission through fuselage panels are extensively investigated in [22, 23]. More recently, Lefevre et al. [24] developed a fully coupled electro-elasto-acoustic model for active noise reduction. Numerical investigations were performed on a smart rectangular plate mounting four piezoelectric patches, used as two collocated sensor/actuator pairs, to control both structural vibrations and radiated sound field. An electroacoustic transducer with tunable mechanical and electrical characteristics was proposed by Breitbach et al. for active control of sound sources [25]. A semi-passive technique based on switched shunted piezoelectric elements was presented by Guyomar et al. [26] and Larbi et al. [27]. In a semi-active piezoelectric shunt damping system, the electrical energy captured across the smart material is dissipated through tunable electrically shunted circuits in order to decrease the overall mechanical energy of the controlled structure by taking advantage of the electromechanical coupling capabilities [28–30].

In general, the noise reduction technologies may be grouped into two main approaches: passive and active methods. Passive noise control enables efficient sound absorption at frequencies above 1 kHz. In contrast, it becomes impractical for low frequency applications due to the weight penalties.

Active vibro-acoustic control is one of the areas that can exploit the emerging smart structures technology to its full extent. The field of active noise control is divided into two major categories: active structural acoustic control (ASAC) and active noise control (ANC). A more detailed review of the two control strategies is covered in Section 1.2. Since the large amount of literature available on this topic, the main emphasis is given to the most relevant aircraft applications.

1.2 Active Noise Control and Active Structural Acoustic Control

Active control solutions may be classified into two groups depending on whether the control objective invokes a structural metric or an acoustic metric. Pure active vibration control utilizes structural sensors, actuators and a control unit to reduce structural vibrations. Active noise control suppresses the disturbance noise field by injecting sound or vibration through structural actuators or speaker arrangements.

Active noise control can thus be categorized depending on the anti-noise strategy used to destructively counteract the disturbance field: (i) at the location of noise reception (ANC) or (ii) at the origin of sound radiation (ASAC).

The first approach is the loudspeaker-based ANC. It consists of anti-noise sound sources driven to cancel sound pressure disturbance locally, thus creating a quiet area around error microphones. The fundamental principle is the destructive wave interference between the primary (disturbance) sound source and the secondary (control) sound source. The primary sound field is cancelled by secondary sources which generate an inverse noise field, usually referred to as "anti-noise", of equal amplitude and opposite phase with respect to the primary or unwanted noise. This is achieved by monitoring the resulting sound field with error microphones properly distributed throughout the controlled space. Since this approach does not affect the origination of sound, it may require several anti-noise sources in order to achieve global noise reduction in large enclosures.

The second approach is referred to as ASAC. It takes advantage of the correlation between structural vibration and radiated sound field in order to reduce sound emanating from a source. The active control system changes the physical characteristics of the overall acoustical system such as the input impedance presented to an external disturbance, the impedance of modes or the nature of boundary conditions. One of the most effective means to minimize sound power radiated from a vibrating structure is to change its vibro-acoustic response into a "weak radiator". A strong radiator moves a large amount of air as it vibrates. It is thus an efficient noise producer. A weak radiator is a vibration mode which radiates sound very inefficiently due to the volume velocity cancellation. Active control can therefore be employed to force the structure to take on the weak radiator mode shapes. By intervening in the structural vibration, secondary inputs can be driven so that the most efficient radiation modes are suppressed or restructured. The structural actuators may be either piezoceramic patches or inertial shakers aiming at modifying the structural vibrations and, consequently, the way the structure radiates sound. The sensors may be either structural or acoustic sensors, distributed throughout the acoustic cavity.

Given either perspective, there are two broad strategies for implementing active control: feedback and feedforward control. In either case, the

1.2. ACTIVE NOISE CONTROL

controller is a filter which transforms either the residual noise signal or the reference signal to be regulated into an actuator drive signal.

Laminar piezo patches bonded to aircraft interior trim panels represent one of the most covered solutions in the field of active structural acoustic control. Collocated actuators and sensors with a local velocity feedback control loop provide a stable active damping solution for reducing the amplitudes of vibration modes. Controlling fuselage trim panels in such a way as to reduce near field radiation is usually referred to as active boundary control (ABC) [31]. The most interesting aspect is that suppressing structural vibrations by cancelling the out-of-plane velocities does not necessarily guarantee a reduction in the radiated sound power. This is because the control forces may, in principle, excite some modes, previously unexcited, which are more efficient radiators of sound. Consequently, the total sound power output may be greater than in the absence of control [32].

ASAC has an advantage over directly controlling the interior noise field in that global far-field attenuation can be achieved by using fewer control actuators [33–38]. Moreover, a much lower system complexity is required, thus increasing its practical relevance [39–41].

1.2.1 *Active noise control applications in aircraft*

The earliest studies on passenger comfort can be found in a textbook by Arnhym reporting that *comfortization is the art of minimizing or eliminating the discomforts inherent to flying* [42, p. 9]. Since such pioneering studies, the importance of a quiet aircraft cabin has been well understood. However, although since Lueg's first application a considerable amount of research has been conducted, practically successful ANC systems for enhanced passenger comfort have still to find wide practical applications in aircraft cabin. This is documented in a number of papers dealing with active control of cabin sound field.

Active noise control of propeller-induced passenger cabin noise was successfully experimented by Elliott *et al.* and presented in [43]. Sixteen loudspeakers were positioned in front of the seats and in the overhead luggage racks of a British Aerospace 748 twin turboprop aircraft to reduce the noise perception of passengers. Thirty-two error microphones were distributed throughout the cabin interior. In-flight experiments showed a reduction of 13 dB at the fundamental BPF of 88 Hz. The attenuation of the second and third harmonic was improved by 12 dB by concentrating the majority of the loudspeakers in the plane of the propellers.

Fuller et al. [44] used piezoelectric actuators bonded to the aft section of the interior fuselage skin of a Cessna business jet to reduce internal noise by feedback control. A multi-channel control algorithm was developed to achieve a global attenuation of about 10 dB at four error microphones positioned at the seat headrests of the aircraft.

Lecce et al. [45] developed a noise and vibration control system using magnetostrictive actuators with the goal of controlling noise and vibration in a turboprop aircraft at frequencies below 500 Hz. Such a system, made up of 42 actuation/sensing devices mounted on frames and stiffeners of the fuselage section, was tested on a full scale fuselage mock-up of an ATR72 aircraft. A genetic algorithm was developed to optimize actuator positions simulated as point forces acting on selected nodes of the FE (finite element) model. A noise reduction of more than 25 dB was achieved for a broadband noise disturbance derived from in-flight experiments.

In contrast to the large number of published works on active noise control, however, practical ANC systems are still only used sporadically in modern aircraft. The huge expectations have led to a large number of patents but most of the technology demonstrators manufactured in laboratory scale are not yet reliable enough to be widely accepted by then aerospace industry.

Active noise and vibration control systems are often concerned with secondary structures, such as interior trim panels, rather than with the primary load-carrying airframe components, as structural control may cause fatigue damage. Placing actuators on non-critical structures yields local control of cabin noise field. On the other hand, massive inertial shakers may be required to globally control the structure by direct-force actuators integrated in the airframe. As a result, because of the complex, heavy and space-consuming additional components, there is still a great demand for novel solutions for interior noise control.

The current active noise control applications in aircraft cabin range from ANC systems for sound field cancellation to active mountings of engine nacelles. Both solutions are particularly effective at low frequencies. On the Saab 2000 turboprop aircraft, a total of 32 loudspeakers mounted on the trim panels and at foot level are used as secondary sources to mitigate the propeller-induced noise up to 500 Hz. An array of 72 microphones are positioned throughout the interior cabin as error sensors. The disturbance is referenced by the controller through synchronization signals taken directly from the propeller shafts. The ANC system reduces the tonal noise by 8 to 14 dB at the fundamental BPF [46]. Similar solutions are nowadays in

1.2. ACTIVE NOISE CONTROL

service on Bombardier Q-Series Dash 8, Raytheon Beech King Air 350, and are optional for other turboprop and business jets.

Active tuned vibration absorbers (ATVAs) attached to the engine pylon are used in most of the aforementioned turbopropeller regional aircraft [47]. Successful experiments have been conducted on a full-scale Bombardier Dash 8 turbopropeller aircraft by using piezoelectric actuators connected with an appropriate active control unit to reduce engine-induced panel vibrations [48]. The adaptive feedforward controller has been able to reduce the interior noise by approximately 28 dB at the blade passage frequency of 61 Hz. Miniature electromagnetic shakers fixed as structural actuators on the frame-stringer intersection have proven to be more effective than ANC solutions at very low frequencies where the fuselage structural modes are well coupled with the cavity modes [49]. A system based on inertial force actuators mounted on the primary structure for controlling sound radiation is being developed for the Airbus A400M [50]. For military aircraft, active mounting of electronic components are decisive to protect sensitive electronic, electrical and mechanical equipment from high impact and vibrational loads.

Interest in reducing interior noise of transportation vehicles has motivated growing research efforts in low-frequency structural-acoustics. Sound transmission into aircraft cabins is very sensitive to the characteristics of certain parts of the double-walled fuselage panels. The overall acoustic properties depend upon the sound transmission loss capabilities of the whole vibro-acoustic system comprising the outer fuselage, fiberglass blankets, interior trim panels and multi-pane windows. As a result, each contribution to the overall interior noise field shall be accurately investigated in order to identify major noise radiating areas. Although the difficulties in quantifying the transmission paths by which acoustic energy reaches the cabin interior with the necessary spatial resolution, extensive full-scale ground and flight measurements have contributed to characterizing the amount of noise entering from aircraft substructures [51].

Interior noise experiments have demonstrated that sound transmission through aircraft windows has a great impact on the cabin interior acoustic field. Although windows may comprise only five to ten percent of the sidewall area, they nevertheless represent a privileged path for structure-borne and airborne noise transmission into the cabin [52–55]. Sound transmission loss tests on fuselage panels, including windows, have shown that the total sound intensity is mainly radiated by the window area and partially absorbed by the trim panel [18]. In-flight experiments using both double layer arrays of microphones and two-microphone sound intensity probes have produced consistent evidence that conventional windows are an important path by which the incoming sound energy enters the aircraft cabin [53].

The weak link in the chain of structural elements and passive treatments protecting aircraft interior has also been confirmed by the high variability in the sound pressure levels measured next to the aircraft windows and usually being the highest at most frequencies. These noise levels are considered highly annoying to a large percentage of passengers, as confirmed by psychoacoustic tests of comfort perception [56]. Thus, although the importance of passenger's protection against noise is often underestimated, the inefficiency of current noise control treatments helps motivate the research interest in novel solutions for controlling the low frequency sound transmission through aircraft panels. However, it should be mentioned that although active noise and vibration control is a quite mature technology, for noise above 1 kHz passive methods still remain the most convenient solution due to the reduced system complexity and power consumption. Nevertheless, active and passive technologies are complementary approaches and their combination can lead to enhanced performance over a wider frequency range.

1.3 Book Overview

This book provides a bridge for the gap existing between robust control theory and practical active noise control systems. It deals with the development, validation and application, through theoretical and experimental research, of active noise control concepts for enhanced passenger vibroacoustic comfort. This topic covers a wide range of aspects in engineering, varying from finite element modelling to practical implementation of adaptive feedforward control techniques, usually dealt with separately in the literature.

The control strategies addressed in this book for active reduction of sound radiation and transmission through panels are tested and discussed along with recommended practices for noise control experiments. Both benefits and limitations in using structural actuation to control sound power radiation of aeronautical structures are highlighted. To this aim, a smart panels system, referred to as "smart window test-bed" throughout the content of this book, has been developed, tested and used as a benchmark for theoretical comparisons. All this information will be useful

1.3. BOOK OVERVIEW

to encourage readers to use these methods for solving their own problems. Such a reference structure has been dynamically complex enough to gain the full appreciation of the current challenges in active noise control applications involving structural-acoustic coupling between vibrating bodies in a bounded medium, while being simple enough to allow an analytical description of the structural-acoustic dynamics, which is in general not an easy task in more complex applications. Its functionality and performance have been successfully demonstrated by real-time measurements aiming at evaluating not only the overall attenuation in terms of radiated sound power and structural vibration reduction but also the potential benefits on passenger vibro-acoustic comfort, as estimated by a Virtual Passenger Model.

This book is organized in six chapters.

Chapter 1 places the work into the context of active noise control applications for enhanced cabin comfort. A literature survey on aircraft interior noise and active control technologies is provided and the main objectives of the book are outlined.

Chapter 2 introduces the basic equations governing the mutual fluid–structure interaction of fully coupled and uncoupled structural and acoustic subsystems. A weak coupling approach for the exterior acoustic radiation problem of a vibrating body is assumed. For simple structures such as baffled panels, the radiation problem is greatly simplified due to the geometry. The Rayleigh integral completely describes the radiated sound pressure resulting from a planar vibrating source in an infinite baffle. Sound power is determined by the normal velocity distribution over the radiating surface of an array of discrete elemental radiators. Finally, the fully coupled FEM/FEM-based (finite element method) solution for structure and fluid is briefly reviewed. The Eulerian unsymmetric displacement/pressure formulation is derived.

Chapter 3 describes the basic principles underlying the sound transmission through multi-panel partitions. A multi-wall structure consists of a set of panels coupled by enclosed fluid domains. The structural and acoustic models are combined by using the modal coupling theory. A strategy for predicting the sound transmission loss properties of multi-panel partitions is presented. In the low frequency region, coupled structural-acoustical resonances, together with mass–air–mass phenomena, impact on the structural transmission loss capabilities. This influence is intended to be compensated by means of active structural acoustic control. Several examples of double- and triple-panel partitions are numerically examined. Based on the simulation results, the most appropriate concept is finally realized in a proof-of-concept study and tested experimentally.

Chapter 4 presents an overview of the general principles and technology of active noise control and reviews examples of practical applications. Feedforward and feedback control approaches are briefly discussed while main focus is given to the design of adaptive feedforward control schemes. Although the error sensors feed back the residual signals to the controller, these algorithms are still termed as feedforward, but they are not open-loop as in pure feedforward systems. Numerical investigations of active control of sound transmission through smart panels controlled by piezoelectric actuators are presented. A number of control microphones and actuators are used while the system is tested against artificially generated noise excitations. The concept of feedforward adaptive control is developed to reduce acoustic radiation by minimizing the sum of the squares of a number of sensor outputs. It is based on a reference signal highly correlated with the disturbance. The magnitude and phase of the control inputs to the piezoelectric actuators are actively controlled. Adaptive multi-channel algorithms are developed to perform this task, assuming that digital filters are used to implement the sampled controller. Multi-input/multi-output controllers are developed by using three control algorithms, namely the multi-channel filtered-X LMS (MELMS), the least maximum mean squares (LMMS) and the scanning error LMS algorithm. Noise simulations are performed by varying the number of piezo actuators and error sensors used for control.

In Chapter 5, an extensive literature survey on active control of sound transmission through multi-panel partitions is reported. Two different control strategies are detailed. Panel control uses vibration actuators mounted on the radiating panel to achieve a global reduction in the sound radiation. Cavity control aims at controlling the acoustic field in the air-gap between the panels by using acoustic actuators. A technological demonstrator of smart structure with structurally integrated piezoceramic actuators is designed, manufactured and tested. Since the development of piezoceramic materials is a very dynamic research field, the market has been intensively screened to identify commercially available piezoelectric actuators suitable for noise and vibration control of transparent panels. A structural-acoustic FE model, in conjunction with an experimentally identified sound transmission model predicting the sound reduction index of the developed smart structure, are presented.

Chapter 6 presents the practical implementation of the developed active noise control system. A suitable control strategy is designed and successfully implemented on a DSP control board in order to perform real-time experiments of active control of sound transmission. Laboratory experiments

1.3. BOOK OVERVIEW

are executed in a reference sound transmission suite, often encountered in noise control engineering. The low-frequency modal properties of the vibro-acoustic facility are investigated. A multi-channel filtered-X version of the adaptive LMS algorithm is developed to derive the optimal control voltage for the piezoelectric actuators. For a given cost function based on the microphone instantaneous outputs, the gradient descent algorithm is developed in MATLAB®/Simulink® and implemented on a dSPACE control board. Both the structural vibration and the sound power radiated from the smart window test-bed are real-time monitored in order to adaptively minimize the effects of the incident noise field. Finally, a qualitative evaluation of the potential comfort benefits for passenger sitting next to aircraft windows is conducted by a human response model-based comfort analysis.

Chapter 2

Fundamentals in Structural Acoustics

2.1 Introduction

In many areas of everyday engineering, the mechanical behaviour of flexible structures interacting with a surrounding fluid plays an important role. The structural-acoustic coupling is relevant to the vibro-acoustic analysis of a wide variety of physical systems and characterizes numerous applications in automotive and aerospace industries. Aircraft interior noise is a typical example.

In the broadest sense, classes of problems dealing with structural-acoustic coupling involve both the dynamics of structures in contact with an enclosed acoustic fluid and the sound radiation from vibrating structures in an exterior acoustic fluid. When the fluid–structure interaction is limited to an interior problem, the internal acoustic pressure influences the structural motion along the fluid–structure interface while, at the same time, the structural velocity field induces pressure fluctuations in the surrounding medium. Phenomena of this kind have to be considered if the fluid of interest is dense or when the structure is highly flexible. Structural and acoustic subdomains are regarded as parts of one coupled system.

The mutual structural–acoustic interaction depends upon the properties of the uncoupled structural and acoustic subsystems. In aircraft cabins, low order structural modes of the sidewall panels couple efficiently to the axial acoustic modes of the cabin volume by displacing and compressing the fluid

18 CHAPTER 2. FUNDAMENTALS IN STRUCTURAL ACOUSTICS

in contact with the structural surface, thus increasing the overall acoustic response in the cavity [57].

When a flexible structure is in contact with a low-density fluid, the acoustic and structural problems can be solved separately by assuming a weak vibro-acoustic coupling. This simulation approach involves modelling the structure as though it were in a vacuum by neglecting any mutual interaction with the fluid. The resulting one-way interaction can be addressed by means of a two-step procedure. The first step consists of calculating the structural vibrations, induced by a given external force, without taking into account the acoustic pressure loading. In the second step, the resulting sound radiation is modelled by considering the structural vibrations as velocity boundary conditions for the acoustic domain.

Real-life structures are often difficult to solve in a closed analytical form. In most cases, geometry and boundary conditions may be complex and numerical methods are required to solve the coupled governing equations for the acoustic and structural domains. Modelling techniques can be classified as either deterministic or statistical methods. In acoustics, the finite element method (FEM) is an established, well-defined tool for investigating interior vibro-acoustic problems [58, 59]. Alternatively, the boundary element method (BEM) is more effective in solving vibro-acoustic systems involving both bounded and unbounded acoustic domains. It offers many computational advantages over FEM as it involves discretizing only the boundary of the domain [60]. Suitable combinations are available by coupling a finite element model of the structure to a finite element model of the fluid (FE-FE), or alternatively, by coupling a finite element model of the structure to a boundary element model of the fluid (FE-BE). FE-FE formulations are most applicable for use with interior acoustic problems. FE-BE formulations are best suited to large-scale problems of acoustic radiation from arbitrarily shaped vibrating bodies in an infinite exterior region [61]. Unbounded exterior acoustic domains can also be solved by FE-FE methods, but they require the use of infinite elements that artificially damp outgoing waves in the far-field in order to meet Sommerfeld's radiation condition.

However, the use of deterministic approaches is restricted to a limited frequency range. At high frequencies, FEM and BEM become inefficient. The statistical energy analysis (SEA) is more suited to high-frequency problems with broadband excitation. A review of the analytical methods and the numerical approaches for modelling and analyzing structural-acoustic systems is given by Atalla and Bernhard [62]. The use of the finite element method for the study of vehicle interior noise is reviewed by Nefske *et al.* [63]. BEM applications in acoustics are addressed in [60]. The modelling of active structural acoustic control problems in the field of active noise control is provided in Lefevre *et al.* [64]. An original finite element formulation for piezoelectric structural acoustic coupled systems is presented by Deu *et al.* [65]. In [66], Vorgote *et al.* propose a hybrid approach for the analysis of vibro-acoustic systems in the mid- and high-frequency range by coupling deterministic methods with statistical energy analysis.

A FEM-based simulation procedure is used throughout this book to model the fully coupled vibro-acoustic response of multi-wall partitions in the frequency range up to 500 Hz. The systems consist of vibrating panels coupled with an enclosed fluid domain. A weak coupling with the exterior fluid is then assumed to predict the resulting sound radiation in an infinite acoustic medium. The structural-acoustic analysis is accomplished with the FEM-software ABAQUS® and the numerical tool MATLAB® is used for pre- and post-processing of the acoustic results.

This chapter presents a survey of the theorethical background of the dynamic interaction between an acoustic fluid and an elastic structure. A brief description of the most common acoustic quantities used to evaluate sound radiation properties and the attenuation capabilities of structures is provided first. Afterwards, the experimental techniques for calculating sound transmission loss of structures are presented. Next, the mathematical foundation of structure-acoustic problems is described with special emphasis on the fluid domain and the coupling boundary conditions. The free field sound radiation associated with the stuctural vibration of radiating panels is also discussed. The acoustic radiation from individual structural modes is detailed, as well as the sound power radiation from baffled vibrating plates.

2.2 Metrics of Sound Transmission

Different quantitative metrics can be found in the literature to evaluate the performance of acoustic barriers by the amount of noise attenuated or transmitted through them. In this section, the most common metrics involving sound transmission through structures are introduced. They are: sound pressure level (SPL), insertion loss (IL) and transmission loss (TL). A typical transmission loss facility for testing the sound insulation properties of specimens is schematically shown in Fig. 2.1. For this purpose, the

20 CHAPTER 2. FUNDAMENTALS IN STRUCTURAL ACOUSTICS

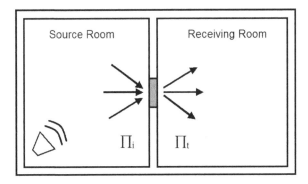

Figure 2.1. Sketch of a transmission loss facility

specimen is mounted in the opening between a source room and a receiving room. This experimental approach is detailed in Section 2.3.

2.2.1 *Sound pressure level*

A preliminary insight on the effectiveness of acoustic barriers in attenuating an incident sound field can be obtained by the difference in the root-mean-square measurements taken at a given position in the receiving room. The sound pressure level, commonly expressed in a logarithmic scale by the dB unit, is defined as

$$\text{SPL} = 20 \log \left(\frac{p_{rms}}{p_0} \right) \tag{2.1}$$

where the reference pressure p_0 is $20\,\mu Pa$ corresponding to the threshold of human hearing at 1 kHz. According to this expression, by doubling the sound pressure, the SPL increases by 6 dB.

The passive attenuation provided by a structural partition can be calculated by

$$\text{SPL}_{attenuation} = 20 \log \left(\frac{p_{rms,\ without\ partition}}{p_{rms,\ with\ partition}} \right). \tag{2.2}$$

This metric must be interpreted carefully when estimating the level of sound attenuation. This is because the transmitted acoustic field depends upon the sound power of the generating source and its directivity, as well as the distance between the source and the receiver. Moreover, the acoustic mode shapes, the cavity absorption and boundary conditions in the receiving room may locally influence the pressure at the receiver position.

2.2. METRICS OF SOUND TRANSMISSION

As a result, it may happen that sound pressure is locally reduced in some field points while increased in some others. This may occur in many cases, especially while using active noise control systems. If the receiving space simulates a free field environment, as in the case of an anechoic room, the sound waves propagate without any reflection. Only in that case, the mean-square sound pressure level varies inversely with the square of the distance from the source. For laboratory measurement purposes, the free field condition is achieved from a certain frequency onwards inside an anechoic chamber by treating the room side walls with sound absorbing materials so that acoustic reflections are negligible.

Nevertheless, more suitable indicators can be used to characterize the acoustic performance of panel partitions by taking into account global measurements. They are: sound insertion loss (IL) and sound transmission loss (TL).

2.2.2 *Sound insertion loss*

Sound insertion loss (IL) of a structure is defined as the difference between two time-averaged sound power measurements in the receiving room due to a noise source in the source room with and without the structure mounted between the two testing rooms. This index provides a more accurate way to evaluate the achieved sound attenuation since it involves the measurement of the transmitted sound power. This global masurement takes into account the total acoustic energy which flows through the partition into the receiving space. The IL in a receving room is expressed as

$$IL = 10 \log \left[\frac{\Pi_{t,\ without\ partition}}{\Pi_{t,\ with\ partition}} \right]. \quad (2.3)$$

By installing the partition in the opening between the source and the receiving room, the amout of energy which is dissipated through the partition is proportional to the transmitted sound power Π_t.

Additionally, insertion loss enables the evaluation of the acoustic benefits in using multi-panel configurations. The insertion loss of a double panel partition is defined as the difference between the transmitted sound power of a double panel and that of a single panel obtained by removing one of the plates. In this case, it is defined as

$$IL = 10 \log \left[\frac{\Pi_{t,\ single}}{\Pi_{t,\ double}} \right]. \quad (2.4)$$

2.2.3 Sound transmission loss

Sound transmission loss (TL) is defined as the amount of sound reduction that a structural barrier imparts to a transmitted sound wave. It is defined as

$$TL = 10\log\left(\frac{1}{\tau}\right) = 10\log\left(\frac{\Pi_i}{\Pi_t}\right) \quad (2.5)$$

where τ is the sound transmission coefficient and Π_i and Π_t are the incident and the transmitted sound powers, respectively.

Figure 2.2 shows a typical sound transmission loss curve for a single panel partition. Sound transmission loss is a frequency selective phenomenon. At low frequencies, it is controlled by the stiffness and by the structural resonances. Above these frequencies, the sound transmission loss strictly depends on the mass law. At higher frequencies, the sound transmission loss properties rapidly decrease due to the coincidence effect. The coincidence frequency is defined as the frequency at which the bending wave speed in the material equals the speed of the sound in air. At this specific frequency, the acoustic wavelength of the sound wave coincides with the wavelength of the flexural wave in the structure so that the sound wave is

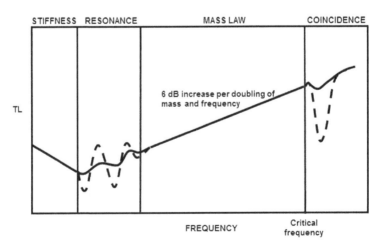

Figure 2.2. Generalized TL spectrum for a typical homogeneous partition [69]

transmitted through the plate without any losses. At the coincidence, the critical frequency equals

$$f_{crit} = \frac{c^2}{2\pi}\sqrt{\frac{\rho_s h}{D}} \qquad (2.6)$$

where D is the bending stiffness of the panel.

2.3 Transmission Loss Experiments

From an acoustic point of view, structures can be considered as frequency-dependent acoustic filters whose effectiveness in attenuating a disturbing sound field depends upon their intrinsic properties. Transmission loss (IL) experiments aim at characterizing such capabilities by quantifying the amount of noise which is attenuated and transmitted through the structure.

Laboratory tests of TL are typically conducted by using either two reverberation chambers or a reverberation chamber coupled with an anechoic chamber. In the former approach, TL is calculated from sound pressure measurements. The latter approach uses sound intensity-based power measurements. The measurement unit is decibel (dB), usually in octave or 1/3 octave bands. The frequency of interest typically ranges from 60 Hz for large chambers to 10 kHz.

A typical experimental layout for TL measurements is shown in Fig. 2.1. For this purpose, the source room is a reverberation chamber. It is characterized by highly reflective side walls which contribute to achieving diffuse sound fields where pressure levels do not depend on the measurement position. The diffuseness of a reverberation chamber increases with the dimensions. The room volume determines the maximum wavelength, or the minimum frequency, for which the sound field can be considered diffuse. This determines the frequency range of use for the facility.

The receiving room may be either a reverberation chamber or an anechoic chamber. The latter is used to experimentally reproduce the free field radiation conditions by suppressing any reflection from the side walls.

The pressure-based method is used to characterize the sound transmission loss of specimens when mounted between two adjacent reverberation rooms. The specimen is mounted in the opening between the two rooms and a diffuse sound field is generated in the source room by an acoustic source emitting a broadband signal in the frequency range of interest. In the hypothesis of perfectly diffuse sound field, the acoustic intensity per unit area at the incident side can be written as

$$I_i = \frac{<p_i^2>}{4\rho_0 c}. \tag{2.7}$$

The transmitted sound power due to the structure-borne noise of the partition is thus

$$\Pi_t = \tau A \frac{<p_i^2>}{4\rho_0 c} \tag{2.8}$$

where τ is the transmission coefficient of the partition of area A.

By taking into account the acoustic absorption in the receiving chamber, TL can be computed by the difference in the space/time averaged sound pressure level between the source and the receving room. According to [70], TL can be expressed as

$$TL = Lp_{source} - Lp_{receiving} + 10 \log \frac{A}{\kappa} \tag{2.9}$$

where Lp_{source} is the space/time averaged sound pressure level in the source room, $Lp_{receiving}$ is the space/time averaged sound pressure level in the receiving room, A is the surface area of the specimen and κ is the total absorption in the receiving chamber. The main uncertainties in the achieved results are related to the possible spatial differences in the sound pressure level at the source room and to the spatial and temporal variations of the reverberation time in the receiving room.

Alternatively, TL properties of structures can be characterized by using the sound intensity-based method. This technique is discussed in more detail in Section 6.6. The specimen is mounted in the opening between a reverberation room and an anechoic room. The sound power transmitted through the structure is directly measured by an intensity probe rather than by a spatial/time averaged measurement of the sound pressure and an estimate of the reverberation time, as for a receiving reverberation room. This approach is less expensive and it is used in many TL facilities. According to [71], TL of the partition is given by

$$TL = Lp_{source} - L_{in} - 6 + 10 \log \frac{A_s}{A_m} \tag{2.10}$$

where Lp_{source} is the space/time average sound pressure level in the source room, L_{in} is the surface averaged transmitted sound intensity normal to the measurement surface, A_s is the area of the test specimen or of a portion of it included in the measurement volume and A_m is the total area of the measurement surface.

2.4 Experimental Modal Analysis

Despite the growing availability of computational power, the possibility of attaining an accurate understanding of structural dynamics by means of numerical methods is still a topic for research. The exploitation of experimental means to validate the numerical predictions still remains the most reliable approach to obtaining an accurate estimation of the actual response of a structure to complex excitations or structural modifications.

Experimental modal analysis is a commonly-used technique to characterize the dynamic behaviour of mechanical structures [72]. It involves a vast range of areas such as instrumentation, signal processing, system identification and vibration analysis. Especially in the low frequency range, the experimental modal approach is used to characterize the vibro-acoustic coupling of structures on the base of vibrational or acoustical frequency response function (FRF) measurements either with respect to acoustical or to structural excitations. The main advantage is that the actual structure, including possible manufacturing anomalies, is tested by making no assumptions on the boundary conditions, such as on the behaviour of joints, but simply measuring the way they contribute to the structural or the acoustic response for a given excitation.

The goal of modal analysis is to determine natural frequencies, normal-mode shapes and modal parameters of a structure over a specified frequency bandwidth. These modal properties can be identified by modelling or by testing. In the latter case, there are several methods for extracting modal data. They are essentially based on two main concepts: the physical separation of modes by sinusoidal excitations or the mathematical separation of modes by broadband global excitations.

The global excitation method, also known as phase separation technique, uses single or multiple broadband excitation signals to excite the structure. The resulting FRFs include all the modes excited simultaneously in the frequency range of interest. Curve fitting methods are then applied to identify the modal data from the measured FRFs.

Alternatively, the force appropriation method, also referred to as normal mode testing, is a common practice in the aerospace industry to measure undamped natural frequencies and the corresponding mode shapes by using multiple exciters with a sinusoidal excitation [73, 74]. Normal modes are inherent characteristic of structures, depending on the material properties (mass, damping and stiffness) and the boundary conditions. In order to isolate a normal mode, a distribution of coherent sinusoidal forces must be applied to the structure at a resonant frequency. As a result, the modal appropriation is achieved physically rather than mathematically by taking advantage of the spatial filtering provided by the orthogonal mode shapes of a structure. They are purely excited when all responses are in monophase and in quadrature with the sinusoidal excitation and the external forces compensate for internal damping [75].

2.4.1 Basics in structural dynamics

The equations of motion of a linear, elastic structure with viscous-type damping are given in matrix notation by

$$\mathbf{M}\ddot{\mathbf{u}}(t) + \mathbf{C}\dot{\mathbf{u}}(t) + \mathbf{K}\mathbf{u}(t) = \mathbf{f}(t). \tag{2.11}$$

$\mathbf{M}$, $\mathbf{C}$ and $\mathbf{K}$ are the mass, damping and stiffness symmetric matrices, respectively. $\mathbf{f}(t)$ is the time-dependent excitation vector and $\mathbf{u}(t)$ is the vector of physical displacements.

Assuming a harmonic excitation force vector $\mathbf{f}(t) = \mathbf{F}e^{j\omega t}$ and the response vector $\mathbf{u}(t) = \mathbf{U}e^{j\omega t}$, the following is obtained

$$\mathbf{U} = \left[\mathbf{K} - \omega^2\mathbf{M} + j\omega\mathbf{C}\right]^{-1}\mathbf{F} = \mathbf{H}(\omega)\mathbf{F} \tag{2.12}$$

where $\mathbf{H}(\omega)$ is the transfer function matrix (receptance) including the system dynamic characteristics.

The governing equations of motion in the frequency domain can be solved by inverting the matrix term $\left[\mathbf{K} - \omega^2\mathbf{M} + j\omega\mathbf{C}\right]$ for each forcing frequency. Each element α_{ik} of the matrix corresponds to an individual FRF, depending on the relation between the response at coordinate i and a force excitation applied at coordinate k.

The eigenvalue problem

The solution for a free, undamped vibration is obtained by solving the eigenvalue problem

$$\left(\mathbf{K} - \omega^2\mathbf{M}\right)\mathbf{U} = 0. \tag{2.13}$$

Equation (2.13) has a non-trivial solution if

$$\det\left(\mathbf{K} - \omega^2\mathbf{M}\right) = 0. \tag{2.14}$$

This algebraic equation, known as the characteristic equation for the system, yields N real solutions, also known as eigenvalues of the system. For each eigenvalue $\lambda_r = \omega_r^2$, the corresponding eigenvector ϕ_r satisfies

$$(\mathbf{K} - \lambda_r\mathbf{M})\phi_r = 0. \tag{2.15}$$

2.4. EXPERIMENTAL MODAL ANALYSIS

A matrix $\boldsymbol{\Phi}$ whose columns contain the orthogonal eigenvectors ϕ_r can be defined, as well as a matrix $\boldsymbol{\Omega}$ containing the eigenfrequencies on the diagonal entries. By performing a modal transformation

$$\mathbf{u}(t) = \sum_{r=1}^{N} \phi_r q_r(t) = \boldsymbol{\Phi}\mathbf{q}(t) \tag{2.16}$$

the structural response is described by the summation of the normal mode shape vectors ϕ_r of the undamped system and generalized coordinates $q_r(t)$. This formulation is also known as modal expansion.

Introducing Eq. (2.16) into Eq. (2.11) and premultiplying with the transponse modal matrix $\boldsymbol{\Phi}^T$ yields the equivalent modal formulation

$$\begin{bmatrix} \ddots & & \\ & \mu & \\ & & \ddots \end{bmatrix} \ddot{\mathbf{q}} + \hat{\mathbf{C}}\dot{\mathbf{q}} + \begin{bmatrix} \ddots & & \\ & \gamma & \\ & & \ddots \end{bmatrix} \mathbf{q} = \boldsymbol{\Phi}^T \mathbf{f} \tag{2.17}$$

expressed in terms of the generalised mass, damping and stiffness matrices. In modal analysis, it is a common practice to scale the mode shape vectors to unity modal mass. In such a case, the orthogonality property reads

$$\boldsymbol{\Phi}^T \mathbf{M} \boldsymbol{\Phi} = \mathbf{I}, \tag{2.18}$$

$$\boldsymbol{\Phi}^T \mathbf{K} \boldsymbol{\Phi} = \begin{bmatrix} \ddots & & \\ & \omega_r^2 & \\ & & \ddots \end{bmatrix}. \tag{2.19}$$

For undamped systems, Eq. (2.17) reduces to a system of N uncoupled equations, each representing a single mass oscillator. The generalized damping matrix $\hat{\mathbf{C}}$ is generally a non-diagonal matrix. Nevertheless, for most practical cases, it is acceptable to assume a viscous damping matrix proportional to the stiffness matrix, to the mass matrix or to a linear combination of both. In such a case,

$$\hat{\mathbf{C}} = \boldsymbol{\Phi}^T \mathbf{C} \boldsymbol{\Phi} = \begin{bmatrix} \ddots & & \\ & 2\xi_r \omega_r & \\ & & \ddots \end{bmatrix}. \tag{2.20}$$

The variable ξ_r is the modal damping parameter for the rth eigenmode and is defined as the ratio between the damping coefficient c_r to the critical damping c_{c_r} characterizing the solution of a single mass oscillator. It establishes the border between an underdamped and an overdamped system.

The introduction of a modal damping matrix corresponds to the assumption that the energy dissipated during the vibration can be considered as the sum of individual dissipation losses for each participating eigenmode.

2.4.2 Modal parameter estimation algorithms

Recent advantages in modal testing methods and in digital signal processing have increased the accuracy and confidence associated with the experimental determination of modal parameters from a set of FRFs. The general FRF term relating the response displacement at coordinate i due to a force excitation at coordinate k is given by [72]

$$\alpha_{ik}(\omega) = \frac{U_i}{F_k} = \sum_{r=1}^{N} \left(\frac{\phi_{ir}\phi_{kr}}{j\omega - s_r} + \frac{\phi_{ir}^*\phi_{kr}^*}{j\omega - s_r^*} \right) \quad (2.21)$$

where $s_r = -\omega_r \xi_r + j\omega_r \sqrt{1-\xi_r^2}$ is the pole value for mode r, $*$ designates complex conjugate and N is the number of modes that contribute to the dynamic response within the frequency range under consideration.

Equation (2.21) can also be expressed in terms of residues as

$$\alpha_{ik}(\omega) = \sum_{r=1}^{N} \left(\frac{{}_rA_{ik}}{\omega_r\xi_r + j\left(\omega - \omega_r\sqrt{1-\xi_r^2}\right)} + \frac{{}_rA_{ik}^*}{\omega_r\xi_r + j\left(\omega - \omega_r\sqrt{1-\xi_r^2}\right)} \right) \quad (2.22)$$

with ${}_rA_{ik}$ residue value for mode r.

Modal extraction methods range from single degree of freedom (SDOF) techniques, such as the peak amplitude method or the circle fitting method, to multiple degree of freedom (MDOF) approaches involving data from multiple excitations and responses. The SDOF methods assume that modes are well separated, damping is not very high and the dynamic response can be well described by the predominant mode in a limited frequency range. The peak amplitude method, for instance, takes the natural frequencies and the damping ratios from the magnitude and sharpness of the FRF at various response locations. However, these methods are very sensitive to the quality of data and are inadequate in the case of closely spaced modes. Among different time domain methods, the complex exponential method (CE) and its MIMO extension, such as the polyreference complex exponential (PRCE) method [76], are widely used modal analysis identification techniques which provide a very accurate modal parameters estimation by fitting the experimental FRF expressed in terms of receptance. On the other hand, the frequency domain methods offer the advantage of considering the residual effects of modes that lie outside the frequency range of analysis [77]. In this book, the commercial software Polymax® implemented on the LMS Test Lab® suite is used to compute the modal parameters of the investigated structure. Such a curve fitter has provided satisfying results in many aeronautical applications and it represents a relevant tool for research environment [78].

2.5 The Acoustic Wave Equation and the Helmholtz Equation

The governing equations in acoustics need to be introduced in order to address interior vibro-acoustic problems. In order to derive the three-dimensional wave equation relating the acoustic variables pressure, density and velocity in the fluid, the principles of mass and momentum conservation are presented first [79]. Consider a homogeneous, inviscid and compressible fluid with a velocity $\mathbf{v}_F$ at any given position in space defined by the vector $\mathbf{x}$ in the three coordinate directions specified by x_1, x_2 and x_3, the balance between the net inflow of mass into a small three-dimensional element and the net increase of mass of the element leads to the mass conservation equation

$$\frac{\partial \rho_F(t)}{\partial t} + \rho_0 \nabla \mathbf{v}_F(t) = 0. \tag{2.23}$$

Similarly, one can relate the acceleration of a given element of fluid to the pressure gradients by applying the principle of momentum conservation to a fluid element. In this case, the independent equations associated with each coordinate direction can be written as

$$\rho_0 \frac{\partial \mathbf{v}_F(t)}{\partial t} + \nabla P(t) = 0. \tag{2.24}$$

The equation that results by taking the divergence of Eq. (2.24) and differentiating Eq. (2.23) with respect to time is

$$\nabla^2 P(t) - \frac{\partial^2 \rho_F(t)}{\partial t^2} = 0. \tag{2.25}$$

Assuming an adiabatic process, synonymous with the relation $\partial P/\partial t = c^2 \partial \rho_F/\partial t$, the governing linear wave equation reads

$$\nabla^2 P(t) - \frac{1}{c^2} \frac{\partial^2 P(t)}{\partial t^2} = 0. \tag{2.26}$$

Note that in rectangular Cartesian coordinates $\nabla^2 P$ is given by the summation

$$\nabla^2 P = \frac{\partial^2 P}{\partial x_1^2} + \frac{\partial^2 P}{\partial x_2^2} + \frac{\partial^2 P}{\partial x_3^2}. \tag{2.27}$$

In case of time harmonic disturbances of frequency ω, for which $P(\mathbf{x}, t) = p(\mathbf{x})e^{j\omega t}$, the resulting wave equation at any point with generalized coordinate $\mathbf{x}$ becomes

$$\nabla^2 p(\mathbf{x}) + k^2 p(\mathbf{x}) = 0 \tag{2.28}$$

where $k = \omega/c$ is the acoustic wave number. This equation is known as the standard Helmholtz equation.

In acoustics, the boundary conditions can be generally divided into the following standard cases:

- A prescribed pressure (Dirichlet boundary condition)

$$p(\mathbf{x}) = \bar{p}(\mathbf{x}), \quad \mathbf{x} \in \Gamma_p \tag{2.29}$$

where $\bar{p}(\mathbf{x})$ is a prescribed pressure function.

- A prescribed normal velocity (Neumann boundary condition)

$$\frac{\partial p(\mathbf{x})}{\partial n} = -\rho_0 \cdot j\omega \cdot \bar{v}_n(\mathbf{x}) \quad \mathbf{x} \in \Gamma_V \tag{2.30}$$

where $\bar{v}_n(\mathbf{x})$ is a prescribed normal velocity function with a unit outward-directed normal to the boundary surface. Note that for acoustically hard surfaces

$$\frac{\partial p(\mathbf{x})}{\partial n} = 0, \quad \mathbf{x} \in \Gamma_0. \tag{2.31}$$

- A prescribed normal impedance (Robin boundary condition)

$$\frac{\partial p(\mathbf{x})}{\partial n} = -\rho_0 \cdot j\omega \cdot \frac{p(\mathbf{x})}{\bar{Z}(\mathbf{x})}, \quad \mathbf{x} \in \Gamma_Z \tag{2.32}$$

where $\bar{Z}(\mathbf{x})$ is the normal impedance function modelling the effect of insulation material.

In free-field coupled vibro-acoustic problems, the pressure field must also satisfy the Sommerfeld radiation condition at infinity, which states that, as the radial distance r from a harmonic source of sound approaches infinity,

$$\lim_{r \to \infty} \{r[p - \rho_0 c v_{r_F}]\} = 0 \tag{2.33}$$

where v_{r_F} is the radial component of the particle velocity field. This condition ensures that at a large radius r from the acoustic source, no reflections occur at the boundary surface at infinity.

2.5. THE ACOUSTIC WAVE EQUATION

2.5.1 The inhomogeneous wave equation

A very useful concept in acoustics is that of a point source. Consider a sphere having a radius a and whose surface pulsates uniformly and harmonically with a complex velocity U_a. Let us assume that the sphere is reduced in radius but the surface velocity is increased such that the product of the surface velocity and the surface area of the sphere remains constant. The source strength q can be defined such that, as $a \to 0$

$$q = U_a 4\pi a^2 = \text{const.} \tag{2.34}$$

The dimensions of the complex source strength are those of volume velocity, i.e. the product of velocity multiplied by surface area.

In a similar way, a source distribution of monopole type can be conceived by dividing a region of the acoustic medium into a number of three-dimensional elements and assuming that each element contains a pulsating sphere which introduces a fluctuating volume flow into the element [80]. This fluctuating flow per unit volume is specified by $q_{vol}(\mathbf{x}, t)$. When the elements in the medium are infinitesimally small, the rate of addition of volume flow is a continuous function. In that case, the mass conservation Eq. (2.23) can be written as

$$\frac{\partial \rho_F(\mathbf{x}, t)}{\partial t} + \rho_0 \nabla \mathbf{v}_F(\mathbf{x}, t) = \rho_0 q_{vol}(\mathbf{x}, t). \tag{2.35}$$

Applying the divergence operator ∇ to the momentum conservation, given by Eq. (2.24), and subtracting the time derivative of Eq. (2.35), the inhomogeneous wave equation takes the form

$$\left(\nabla^2 - \frac{1}{c^2}\frac{\partial^2}{\partial t^2}\right) p(\mathbf{x}, t) = -\frac{\rho_0 \partial q_{vol}(\mathbf{x}, t)}{\partial t} \tag{2.36}$$

where the term on the right side can be interpreted as source of sound. In case of harmonic fluctuations, Eq. (2.36) becomes

$$\nabla^2 p(\mathbf{x}) + k^2 p(\mathbf{x}) = -j\omega \rho_0 q_{vol}(\mathbf{x}) \tag{2.37}$$

which is known as the inhomogeneous Helmholtz equation.

2.5.2 The solution to the inhomogeneous wave equation

Using the coordinate $\mathbf{y}$ instead of the coordinate $\mathbf{x}$ and defining ∇_y as the divergence operator, i.e. $\mathbf{i}\partial/\partial y_1 + \mathbf{j}\partial/\partial y_2 + \mathbf{k}\partial/\partial y_3$, where $\mathbf{i}$, $\mathbf{j}$ and $\mathbf{k}$ are the unit vectors, Eq. (2.37) becomes

$$\nabla_y^2 p(\mathbf{y}) + k^2 p(\mathbf{y}) = -Q_{vol}(\mathbf{y}) \qquad (2.38)$$

where the source strength distribution is

$$Q_{vol}(\mathbf{y}) = j\omega\rho_0 q_{vol}(\mathbf{y}). \qquad (2.39)$$

In order to determine the solution to the inhomogeneous Helmholtz equation in the form of an integral equation for the complex pressure, the Green function is introduced.

Considering an unbounded domain perturbed by a unit concentrated harmonic source at $\mathbf{y}$, the Green function is a solution to the inhomogeneous Helmholtz equation [81]

$$\left(\nabla^2 + k^2\right) G(\mathbf{x}|\mathbf{y}) = -\delta(\mathbf{x} - \mathbf{y}) \qquad (2.40)$$

where $\delta(\mathbf{x} - \mathbf{y})$ is the three-dimensional Dirac delta function representing a point source at $\mathbf{y}$. The Dirac delta function is a generalized function having the following properties

$$\int_V f(\mathbf{x})\delta(\mathbf{x} - \mathbf{y})dV = \begin{cases} f(\mathbf{y}), & \mathbf{y} \text{ within } V \\ \frac{1}{2}f(\mathbf{y}), & \mathbf{y} \text{ on the boundary of } V \\ 0, & \mathbf{y} \text{ outside } V \end{cases} \qquad (2.41)$$

where V is a volume in the medium bounded by the surface S.

$G(\mathbf{x}|\mathbf{y})$ determines the spatial dependence of the complex pressure field produced by a harmonic point monopole source located at $\mathbf{y}$. Consequently, the complex pressure p at $\mathbf{x}$ due to the harmonic point monopole source of unit strength at $\mathbf{y}$ can be written as

$$p(\mathbf{x}) = j\omega\rho_0 G(\mathbf{x}|\mathbf{y}). \qquad (2.42)$$

The Green function satisfies the principle of reciprocity. This is given by

$$G(\mathbf{y}_2|\mathbf{y}_1) = G(\mathbf{y}_1|\mathbf{y}_2). \qquad (2.43)$$

This means that the complex pressure at $\mathbf{y}_2$ produced by a point monopole source at $\mathbf{y}_1$ is identical to the complex pressure at $\mathbf{y}_1$ produced by a point monopole source at $\mathbf{y}_2$.

The solution to the inhomogeneous wave equation given by Eq. (2.38) can be derived by multiplying the equation by $G(\mathbf{y}|\mathbf{x})$ and subtracting Eq. (2.40) multiplied by $p(\mathbf{y})$. This yields;

$$G(\mathbf{y}|\mathbf{x})\nabla_y^2 p(\mathbf{y}) - p(\mathbf{y})\nabla_y^2 G(\mathbf{y}|\mathbf{x}) = -Q_{vol}(\mathbf{y})G(\mathbf{y}|\mathbf{x}) + p(\mathbf{y})\delta(\mathbf{y} - \mathbf{x}). \qquad (2.44)$$

2.5. THE ACOUSTIC WAVE EQUATION

Integrating over a volume V that is bounded by the surface S and using the properties of the delta function, it follows that

$$\int_V (G(\mathbf{y}|\mathbf{x})\nabla_y^2 p(\mathbf{y}) - p(\mathbf{y})\nabla_y^2 G(\mathbf{y}|\mathbf{x}))dV + \int_V Q_{vol}(\mathbf{y})G(\mathbf{y}|\mathbf{x})dV$$
$$= \begin{cases} p(\mathbf{x}), & \mathbf{x} \text{ within } V \\ 0, & \mathbf{x} \text{ outside } V \end{cases}. \qquad (2.45)$$

By applying the Green's theorem and using the reciprocity property $G(\mathbf{y}|\mathbf{x}) = G(\mathbf{x}|\mathbf{y})$, Eq. (2.45) for points $\mathbf{x}$ inside V can be written as

$$p(\mathbf{x}) = \int_V Q_{vol}(\mathbf{y})G(\mathbf{x}|\mathbf{y})dV + \int_S [G(\mathbf{x}|\mathbf{y})\nabla_y p(\mathbf{y}) - p(\mathbf{y})\nabla_y G(\mathbf{x}|\mathbf{y})] \cdot \mathbf{n} dS \qquad (2.46)$$

where $\mathbf{n}$ is the unit vector normal to the surface S pointing outwards from the volume.

The solution given by Eq. (2.46) thus requires the knowledge of the distribution of the source strength $Q_{vol}(\mathbf{y})$ in the volume, the Green function $G(\mathbf{x}|\mathbf{y})$, the pressure $p(\mathbf{y})$ and the pressure gradient $\nabla_y p(\mathbf{y})$ on the surface S.

2.5.3 Free space sound radiation

Consider the bounding surface S of the acoustic volume V to be at infinity. In such a case, the surface integral of Eq. (2.46) is zero due to the Sommerfeld radiation condition. Equation (2.46) thus reduces to

$$p(\mathbf{x}) = \int_V Q_{vol}(\mathbf{y})G(\mathbf{x}|\mathbf{y})dV \qquad (2.47)$$

requiring the knowledge of the acoustic source strength distribution over the volume V. In addition, the Green function takes the form [81]

$$g(\mathbf{x}|\mathbf{y}) = \frac{e^{-jk|\mathbf{x}-\mathbf{y}|}}{4\pi|\mathbf{x}-\mathbf{y}|}. \qquad (2.48)$$

This function, known as the free space Green function or the fundamental solution for the Helmholtz equation, satisfies Eq. (2.40) as well as the Sommerfeld radiation condition

$$\lim_{|\mathbf{x}-\mathbf{y}|\to\infty} \left[|\mathbf{x}-\mathbf{y}| \cdot \left(\frac{\partial g(\mathbf{x}|\mathbf{y})}{\partial |\mathbf{x}-\mathbf{y}|} + jkg(\mathbf{x}|\mathbf{y}) \right) \right] = 0. \qquad (2.49)$$

In case of a point monopole source at $\mathbf{y}_1$ where $q_{vol} = q\delta(\mathbf{y}-\mathbf{y}_1)$, the pressure distribution over the volume V is therefore given by

$$p(\mathbf{x}) = \int_V j\omega\rho_0 q_{vol}(\mathbf{y})g(\mathbf{x}|\mathbf{y})dV = \frac{j\omega\rho_0 q e^{-jk|\mathbf{x}-\mathbf{y}_1|}}{4\pi|\mathbf{x}-\mathbf{y}_1|}. \qquad (2.50)$$

2.5.4 The Kirchoff–Helmholtz integral equation

In the absence of any source strength $Q_{vol}(\mathbf{y})$ within the volume V, the solution to the inhomogeneous wave equation, given by Eq. (2.46), reduces to

$$p(\mathbf{x}) = \int_S [g(\mathbf{x}|\mathbf{y})\nabla_y p(\mathbf{y}) - p(\mathbf{y})\nabla_y g(\mathbf{x}|\mathbf{y})] \cdot \mathbf{n} dS \qquad (2.51)$$

where S is the bounding surface and $p(\mathbf{y})$ and $\nabla_y p(\mathbf{y})$ denote the complex pressure and its gradient on the surface. This equation is known as the Kirchoff–Helmholtz integral equation. For a given surface velocity $\mathbf{v}(\mathbf{y})$, the first term of the surface integral given by Eq. (2.51) can be written as:

$$\int_S g(\mathbf{x}|\mathbf{y})\nabla_y p(\mathbf{y}) \cdot \mathbf{n} dS = -\int_S g(\mathbf{x}|\mathbf{y}) j\omega \rho_0 \mathbf{v}(\mathbf{y}) \cdot \mathbf{n} dS$$

$$= \int_S g(\mathbf{x}|\mathbf{y}) j\omega \rho_0 q_{surf}(\mathbf{y}) \cdot dS \qquad (2.52)$$

where $q_{surf}(\mathbf{y}) = -\mathbf{v}(\mathbf{y}) \cdot \mathbf{n}$ is a volume velocity per unit surface area. Note that the $\mathbf{n}$ points out from the volume considered. The pressure at a given position within the volume V is thus generated by an equivalent monopole source distribution over the bounding surface S. Analogously, it can be demonstrated that the second term of the surface integral given by Eq. (2.51) can be specified by equivalent dipole source layers [80].

2.6 Sound Radiation from Vibrating Flat Panels

In recent years, a considerable amount of research has been devoted to the study of sound radiation from plane surfaces subjected to different kinds of excitation and boundary conditions [83, 84]. In this context, the characteristics of radiation from baffled plates are often used as a basis on which to represent the sound radiation from more complex plate-like structures [85, 86].

The basic equation for the calculation of the sound pressure field in an unbounded domain due to a closed body vibrating at circular frequency ω is expressed by the Kirchoff–Helmholtz equation

$$p(\mathbf{r}, \omega) = \int_S \left[p(\mathbf{r}_s) \frac{\partial g(\mathbf{r}|\mathbf{r}_s)}{\partial n} + j\rho_0 \omega v_n(\mathbf{r}_s, \omega) g(\mathbf{r}|\mathbf{r}_s) \right] dS \qquad (2.53)$$

2.6. SOUND RADIATION FROM VIBRATING FLAT PANELS

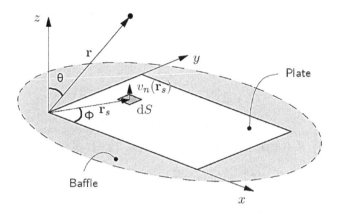

Figure 2.3. Coordinate system of a baffled vibrating plate

where $\mathbf{r}$ is a field point, $v_n(\mathbf{r}_s)$ is the normal complex velocity amplitude at a point $\mathbf{r}_s$ on the structural surface of area S, n denotes the outward normal direction perpendicular to the surface and directed into the fluid, and $g(\mathbf{r}|\mathbf{r}_s)$ is the free space Green function. Such a pressure distribution is equivalent to the sound field due to the superimposition of simple monopole and dipole sources distributed over the surface. For the sake of simplicity, the frequency dependence will be omitted henceforth.

For a planar vibrating plate set in an infinitely extended baffle, as shown in Fig. 2.3, the Green function satisfies the Neumann boundary condition on the surface and it is twice as much as it would be in a free space without the baffle [81]. Therefore, Eq. (2.53) becomes

$$p(\mathbf{r}) = \frac{j\omega\rho_0}{2\pi} \int_S v_n(\mathbf{r}_s) \frac{e^{-jk|\mathbf{r}-\mathbf{r}_s|}}{|\mathbf{r}-\mathbf{r}_s|} dS \qquad (2.54)$$

where $|\mathbf{r} - \mathbf{r}_s|$ is the distance between a field point and a surface point. Equation (2.54) is known as the second Rayleigh integral. Note that it is a two-dimensional convolution of the vibrational velocity and the Green function. The resulting sound pressure depends on the location of the receiver with respect to the sound source.

The assumption that the plate is set in an infinite baffle means that the normal velocity is known on the whole of the plane containing the plate. It is zero outside the surface area. This simplifies the calculations compared to the Kirchoff–Helmholtz equation given by Eq. (2.53). However, analytical solutions are available only for a few cases. For a moving rigid piston, the

36 CHAPTER 2. FUNDAMENTALS IN STRUCTURAL ACOUSTICS

solution is simple and straightforward since the velocity $v_n(\mathbf{r}_s)$ is uniform accross the surface.

In case of simply supported panels, the normal velocity distribution is given by the analytical expression

$$v_n(\mathbf{r}_s) = V_{mn} \sin\left(\frac{m\pi x}{a}\right) \sin\left(\frac{n\pi y}{b}\right) \tag{2.55}$$

for $0 \leq x \leq a$ and $0 \leq y \leq b$. V_{mn} is the complex velocity amplitude of mode (m,n). Substituting Eq. (2.55) into Eq. (2.54) gives

$$p(\mathbf{r}) = \frac{j\omega\rho_0 V_{mn}}{2\pi} \int_0^a \int_0^b \frac{\sin\left(\frac{m\pi x}{a}\right) \sin\left(\frac{n\pi y}{b}\right) e^{-jk|\mathbf{r}-\mathbf{r}_s|}}{|\mathbf{r}-\mathbf{r}_s|} dx\,dy. \tag{2.56}$$

In the far field, where $R = |\mathbf{r} - \mathbf{r}_s|$ is much greater than the source size, defined by the larger edge of the two panel dimensions a and b, Eq. (2.56) becomes [81]

$$p(r,\theta,\phi) = \frac{j\rho_0\omega V_{mn} e^{-jkr}}{2\pi r} \int_0^a \int_0^b \sin\left(\frac{m\pi x}{a}\right) \sin\left(\frac{n\pi y}{b}\right) e^{[j(\frac{\alpha x}{a})+j(\frac{\beta y}{b})]} dx\,dy \tag{2.57}$$

where $r = |\mathbf{r}|$ and

$$\alpha = ka\,\sin\theta\,\cos\phi, \quad \beta = kb\,\sin\theta\,\sin\phi. \tag{2.58}$$

Comparison of Eqs. (2.56) and (2.57) shows that the distance R between a surface element and the observation point is related to the distance r between the observation point and the coordinate origin, by the approximate relationship

$$R \cong r - x\,\sin\theta\,\cos\phi - y\,\sin\theta\,\sin\phi. \tag{2.59}$$

Provided that $R \gg a$ and $R \gg b$, the solution to Eq. (2.57) can be expressed as

$$\tilde{p}_{mn}(r,\theta,\phi) = L_{mn}(\mathbf{r}) V_{mn} \tag{2.60}$$

where

$$L_{mn}(\mathbf{r}) = \frac{jk\rho_0 c}{2\pi}\left(\frac{e^{-jkr}}{r}\right)\Phi \tag{2.61}$$

and

$$\Phi = \frac{ab}{\pi^2 mn}\left[\frac{(-1)^m e^{-j\alpha} - 1}{(\alpha/m\pi)^2 - 1}\right]\left[\frac{(-1)^n e^{-j\beta} - 1}{(\beta/n\pi)^2 - 1}\right]. \tag{2.62}$$

2.6. SOUND RADIATION FROM VIBRATING FLAT PANELS

Equation (2.60) defines the sound pressure at position $\mathbf{r}$ radiated by mode (m,n). Note that $\tilde{p}$ is proportional to $\frac{e^{-jkr}}{r}$. Moreover, it is proportional to the surface of the plate. The sound pressure is inversely proportional to the mode number of the structural mode. The terms α and β determine the sound radiation directivity. An estimate of the velocity can be obtained by considering the first $M \times N$ modes. The sound pressure radiated from the vibrating plate is thus:

$$p(\mathbf{r}) = \sum_{m=1}^{M} \sum_{n=1}^{N} L_{mn}(\mathbf{r}) V_{mn}. \qquad (2.63)$$

2.6.1 Sound power radiation

Two approaches are available in the literature to determine the sound power radiated from a vibrating plate for a given normal velocity distribution over the surface [69]. One is expressed in terms of the amplitude of structural modes and the other in terms of an array of elemental radiators. Both are based on the Rayleigh integral and they assume a weak coupling between the vibrating structure and the radiated sound field.

The first approach, proposed by Wallace [82], uses an expression for the far-field radiated sound pressure and the free-field acoustic impedance for plane waves to derive the sound intensity in the far-field. This sound intensity is then integrated over a sphere enclosing the plate and with large radius, compared to the wavelength of the sound waves at the frequency being considered, to evaluate the radiated sound power.

The second method, introduced by Vitiello et al. [87] and also used by Cunefare et al. [22] and Elliott et al. [88] is based on a near-field approach. It divides the radiating structure into a number of elementary radiators. The sound pressure close to the surface of each of these radiators is given by the sum of the contributions from all the radiators in the entire surface. The computation of the sound power from each radiator is based on the impedance of a point source radiating into the free-field.

Both models yield similar results [88] as they are both based on the free field assumption. The near-field approach, however, is much easier to implement, much less computationally demanding, and particularly suitable for FE models.

The far-field sound intensity of the simply supported baffled finite plate, shown in Fig. 2.3, at a given frequency $\omega = kc$ in a point $\mathbf{r} = (r, \theta, \phi)$ is given by

$$I(\mathbf{r}) = \frac{|p(r,\theta,\phi)|^2}{2\rho_0 c}. \qquad (2.64)$$

The total sound power $\Pi(\omega)$ radiated by the baffled rectangular plate is thus computed by integrating the acoustic intensity over a hemisphere surface:

$$\Pi = \int_0^{2\pi} \int_0^{\pi/2} I(\mathbf{r})\, r^2 \sin\theta\, d\theta\, d\phi. \qquad (2.65)$$

Substituting Eq. (2.63) into Eq. (2.65) gives

$$\Pi = \sum_{m=1}^{M}\sum_{n=1}^{N}\sum_{\acute{m}=1}^{M}\sum_{\acute{n}=1}^{N} \left\{ V_{mn} V_{\acute{m}\acute{n}}^{*} \int_0^{2\pi}\int_0^{\pi/2} \frac{L_{mn}(\mathbf{r}) L_{\acute{m}\acute{n}}^{*}(\mathbf{r})}{2\rho_0 c} r^2 \sin\theta\, d\theta\, d\phi \right\}. \qquad (2.66)$$

Equation (2.66) shows that the total radiated sound power depends on the contribution of a combination of modes, referred to as self-modal radiation for $m = \acute{m}$ and $n = \acute{n}$, and cross-modal radiation for $m \neq \acute{m}$ and $n \neq \acute{n}$.

The acoustic power output at a single frequency can thus be written as

$$\Pi = \dot{\boldsymbol{\eta}}^H \hat{\mathbf{M}} \dot{\boldsymbol{\eta}} \qquad (2.67)$$

in which $\dot{\boldsymbol{\eta}}$ is the structural modal velocity vector and the superscript ^{H} denotes the Hermitian (conjugate transpose). $\hat{\mathbf{M}}$ is the matrix of the structural mode radiation resistances. It is real, symmetric and positive definite. The diagonal terms represent the self-radiation resistance of the structural modes. The off-diagonal terms are the mutual-radiation resistances, representing the interaction between two different structural modes to form the radiated sound power.

For a given velocity distribution on the surface expressed in terms of amplitudes of a finite number of structural modes, the acoustic power radiated by the panel can thus be calculated as the sum of the squared moduli of each structural mode amplitude multiplied by its self-radiation resistance, and the sum of cross products of the structural mode amplitudes multiplied by their mutual radiation resistance. Since $\hat{\mathbf{M}}$ is a fully populated matrix, the off-diagonal terms cannot be neglected in the prediction of the sound power. As a result, active vibration control of a single structural mode does not necessarily imply a reduction in the radiated sound power. In [89], Snyder et al. demonstrate that the cross-modal coupling (non-zero off-diagonal

2.6. SOUND RADIATION FROM VIBRATING FLAT PANELS

elements in $\hat{\mathbf{M}}$) can occur only between a pair of modes with the same class, i.e. (odd-odd) (even-even).

2.6.2 Elemental radiators

The sound power radiated by a baffled vibrating structure can be calculated by applying a direct numerical integration scheme to the Rayleigh integral. This near-field approach is equally valid for both monopole sources in free space and elemental piston sources on the radiating surface. The Rayleigh integral is interpreted in such a way that the sound radiation is the result of a superposition of elemental radiators or pistons that move with constant harmonic velocity. This involves dividing the surface of the panel into a finite number of rectangular elements of equal size, which are small compared to the acoustic wavelength. The normal velocity is assumed constant accross each element. Each elemental source of the plate is a discrete monopole source of sound radiating in the free field. For this discretization, Eq. (2.54) can be written as

$$\mathbf{p} = \mathbf{Z}\mathbf{V}_n \qquad (2.68)$$

where $\mathbf{p}$ is the vector with pressures in a set of field points, $\mathbf{V}_n$ is the vector with normal velocities of the elemental radiators and $\mathbf{Z}$ is a frequency dependent impedance matrix whose elements are given by

$$Z_{ij} = \frac{j\rho_0 \omega S_e}{2\pi} \frac{e^{-jkr_{ij}}}{r_{ij}} \qquad (2.69)$$

where S_e is the area of the elemental radiator and $r_{ij} = |\mathbf{r}_i - \mathbf{r}_j|$ is the distance between a field point i and a surface point j, where $\mathbf{r}_j$ refers to the centre of the elemental radiator.

When dealing with a FE model of a planar radiating surface, the nodes can be assumed as the centres of the elementary radiators. The velocity distribution can be derived from a coupled or an uncoupled vibro-acoustic analysis. The former is particularly indicated for the sound transmission analysis of double and triple panel partitions since it takes into account the vibro-acoustic coupling between the panels and the interior air cavities.

Equation (2.68) can be written in full as

$$\begin{bmatrix} p_1 \\ p_2 \\ \vdots \\ p_n \end{bmatrix} = \frac{j\omega\rho_0 S}{2\pi} \begin{bmatrix} \frac{e^{-jkr_{11}}}{r_{11}} & \frac{e^{-jkr_{12}}}{r_{12}} & \cdots & \frac{e^{-jkr_{1n}}}{r_{1n}} \\ \frac{e^{-jkr_{21}}}{r_{21}} & \frac{e^{-jkr_{22}}}{r_{22}} & \cdots & \frac{e^{-jkr_{2n}}}{r_{2n}} \\ \cdots & \cdots & \cdots & \cdots \\ \frac{e^{-jkr_{n1}}}{r_{n1}} & \frac{e^{-jkr_{n2}}}{r_{n2}} & \cdots & \frac{e^{-jkr_{nn}}}{r_{nn}} \end{bmatrix} \begin{bmatrix} V_1 \\ V_2 \\ \vdots \\ V_n \end{bmatrix} \qquad (2.70)$$

where n is the number of sources. At the source positions ($r = 0$), the evaluation of the pressure response is affected by the singularity associated with the imaginary part of the specific impedance, known as radiation reactance. Such value is typically derived by referring to the acoustic radiation impedance of a piston in a baffle. For a rigid piston of dimensions $a \times b$, where $b \cong a$, moving with uniform velocity V_n, the radiation impedance is given by [90].

$$\frac{p}{V_n} = \frac{\rho_0 c k^2}{16}(a^2 + b^2) + \frac{j 8 \rho_0 c k}{9\pi}\left(\frac{a^2 + ab + b^2}{a+b}\right), \quad ka \ll 1. \tag{2.71}$$

Applying Eq. (2.71) to the elemental source dS so that $a = dx = b = dy$, this reduces to

$$\frac{p}{V_n} = \rho c \left(\frac{k^2 dx^2}{8} + \frac{j 4 k dx}{3\pi}\right), \quad k\, dx \ll 1. \tag{2.72}$$

However, it is in the sound power computation that this method shows the major advantages. The time averaged total sound power radiated by a baffled vibrating plate can be expressed as [91]

$$\Pi = \frac{1}{2}\int_S \Re\{p(\mathbf{r})v_n^*(\mathbf{r}_s)\}\, dS \tag{2.73}$$

where * denotes the complex conjugate.

If the sound pressure is calculated from the discretized representation of the Rayleigh integral, the acoustic power radiated from the array of elemental sources is proportional to the real part of the sum of the conjugate volume velocities of each radiator multiplied by the corresponding acoustic pressure. It can be written as

$$\Pi = \frac{1}{2}\Re\left\{\mathbf{V}^H \mathbf{p}\right\} S \tag{2.74}$$

where S is the sum of the area of the elementary radiators which are assumed to be of equal size for convenience.

The symmetry of the impedance matrix $\mathbf{Z}$ enables to express the sound power in terms of the velocity $\mathbf{V}$ of the elemental radiators, as

$$\Pi = \frac{1}{2}\Re\left\{\mathbf{V}^H \mathbf{Z}\mathbf{V}\right\} S = \mathbf{V}^H \mathbf{R}\mathbf{V} \tag{2.75}$$

where the radiation resistance matrix $\mathbf{R}$ is real, symmetric and positive definite. It is defined as

$$\mathbf{R} = \frac{S}{2}\Re\{\mathbf{Z}\} \tag{2.76}$$

2.6. SOUND RADIATION FROM VIBRATING FLAT PANELS

and for an array of n elemental radiators it can be written in full as

$$\mathbf{R} = \frac{\omega^2 \rho_0 S^2}{4\pi c} \begin{bmatrix} 1 & \frac{\sin(kr_{12})}{kr_{12}} & \cdots & \frac{\sin(kr_{1n})}{kr_{1n}} \\ \frac{\sin(kr_{21})}{kr_{21}} & 1 & \cdots & \frac{\sin(kr_{2n})}{kr_{2n}} \\ \cdots & \cdots & \cdots & \cdots \\ \frac{\sin(kr_{n1})}{kr_{n1}} & \frac{\sin(kr_{n2})}{kr_{n2}} & \cdots & 1 \end{bmatrix}. \quad (2.77)$$

The diagonal terms of the matrix ($i = j \to kR = 0$) tend to unity since $\sin(kR)/kR$ tends to 1 as kR tends to zero.

Using the matrix of radiation resistances of a set of elemental radiators, it is thus possible to compute the sound power radiated into free-space as well as the radiation efficiency of a set of structural modes. This method allows the dynamic response of a radiating body to be considered separately from its sound radiating behaviour being the sound radiation directly expressed in terms of the velocities of a discrete number of points on the radiating surface.

2.6.3 Radiation modes

The eigenvectors of the matrix $\hat{\mathbf{M}}$ of the structural mode radiation resistances are known as radiation modes. They describe a linearly independent velocity distribution of the structure that produces orthogonal sound pressure fields [92]. The corresponding eigenvalues are proportional to the radiation efficiencies and they indicate the contribution of each mode to the total radiated sound power. The radiation modes are calculated from the radiation matrix $\hat{\mathbf{M}}$ as follows

$$\hat{\mathbf{M}} = \mathbf{P}^T \mathbf{\Omega} \mathbf{P} \quad (2.78)$$

where $\mathbf{P}$ is the orthogonal matrix of eigenvectors and $\mathbf{\Omega}$ is a diagonal matrix whose elements Ω_n are the eigenvalues of $\hat{\mathbf{M}}$. The sound power can thus be written as

$$\Pi = \dot{\boldsymbol{\eta}}^H \hat{\mathbf{M}} \dot{\boldsymbol{\eta}} = \dot{\boldsymbol{\eta}}^H \mathbf{P}^H \mathbf{\Omega} \mathbf{P} \dot{\boldsymbol{\eta}}. \quad (2.79)$$

By introducing $\mathbf{b} = \mathbf{P}\dot{\boldsymbol{\eta}}$ which is a vector of transformed radiation mode amplitudes, Eq. (2.79) becomes

$$\Pi = \mathbf{b}^H \mathbf{\Omega} \mathbf{b} = \sum_{n=1}^{N} \Omega_n |b_n|^2. \quad (2.80)$$

The sound power is therefore given by the summation of the contribution of the radiation modes which radiate sound independently. The eigenvectors

Figure 2.4. The two lowest order structural mode shapes of a baffled simply supported panel [81]

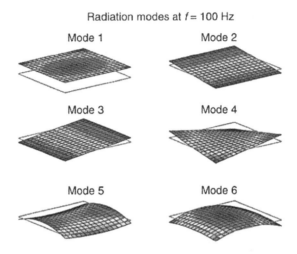

Figure 2.5. Radiation modes at 100 Hz of a baffled simply supported rectangular plate [69]

P can be considered as weighting functions that weight the amplitude of the structural modes to form the radiation modes. In Fig. 2.4 the first two structural mode shapes of a baffled simply supported rectangular plate are schematically depicted. The six lowest order radiation modes at 100 Hz are shown in Fig. 2.5. At low frequencies, the first radiation mode, corresponding to a piston-like mode, is by far the most efficient one. Above 1 kHz, the different modes are generally equally efficient.

2.6.4 *Sound radiation efficiency*

The sound radiation efficiency is a dimensionless measure for how well a vibrating structure radiates sound. It is defined as the ratio of the average acoustic power radiated per unit area by a vibrating surface to the average acoustic power radiated per unit area by a baffled piston which vibrates at a high frequency ($kR \gg 1$, with R the effective piston radius) with a velocity equal to the space and time-averaged, squared velocity $\langle \bar{v}_n^2 \rangle$ of the

2.6. SOUND RADIATION FROM VIBRATING FLAT PANELS

structure [81]. The radiated power from the reference piston is given by

$$\Pi = \rho_0 c S \langle \bar{v}_n^2 \rangle. \tag{2.81}$$

The radiation efficiency is thus

$$\sigma = \frac{\Pi}{\rho_0 c S \langle \bar{v}_n^2 \rangle} \tag{2.82}$$

where S is the total area of the radiating surface. The space and time average of the square normal velocity can be written as

$$\langle \bar{v}_n^2 \rangle = \frac{1}{2S} \int_S |v_n(\mathbf{r}_s)|^2 dS. \tag{2.83}$$

For a given panel deformation shape, the radiation efficiency is typically quantified by analyzing the radiation efficiencies of the structural modes that contribute to that shape. This is particularly important when dealing with sound transmission characteristics of multi-panel partitions since the shape of the panel deformations is a determining factor for the interaction between the incident and transmitted sound fields.

In Fig. 2.6 the radiation efficiencies of the $(1,1)$, $(2,1)$ and $(3,1)$ radiation modes for a simply supported, baffled, rectangular panel are shown as a function of normalized excitation frequency L_x/λ [93]. The aspect ratio of the surface is $L_x/L_y \approx 0.79$. The self-radiation efficiency of the $(1,1)$ structural mode vibrating alone is denoted by $\sigma_{11,11}$ and the mutual radiation efficiency of the $(1,1)$ structural mode and the $(3,1)$ structural mode

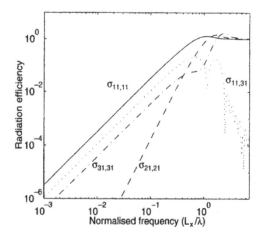

Figure 2.6. Sound radiation efficiencies of the modes of a baffled simply supported panel $((L_x/\lambda)/L_y \approx 0.79)$ when radiating into free space [93]

vibrating together is denoted by $\sigma_{11,31}$. In the low frequency region, the (1, 1) mode and the (3, 1) mode radiate more efficiently than the (2, 1) mode. The (odd-odd) panel modes are identified as efficient radiators, since they produce a net volume displacement of the fluid. This volumetric component creates the predominant sound at low frequencies. The highest values are shown for the piston-like mode.

The radiation efficiency is below unity for frequencies lower than the critical frequency. At the critical frequency, the structural bending wavelength equals the acoustic wavelength and the radiation efficiency exceeds unity. At higher frequencies, it tends to unity.

Approximate expressions for the radiation efficiency of rectangular panels at low frequencies are given by Wallace [81]. They are:

(1) For p and q both odd integers

$$\sigma_{pq} \cong \frac{32(ka)(kb)}{p^2 q^2 \pi^5} \left\{ 1 - \frac{k^2 ab}{12} \left[\left(1 - \frac{8}{(p\pi)^2}\right) \frac{a}{b} + \left(1 - \frac{8}{(q\pi)^2}\right) \frac{b}{a} \right] \right\}.$$
(2.84)

(2) For p odd and q even

$$\sigma_{pq} \cong \frac{8(ka)(kb)^3}{3p^2 q^2 \pi^5} \left\{ 1 - \frac{k^2 ab}{20} \left[\left(1 - \frac{8}{(p\pi)^2}\right) \frac{a}{b} + \left(1 - \frac{24}{(q\pi)^2}\right) \frac{b}{a} \right] \right\}.$$
(2.85)

(3) For p and q both even integers

$$\sigma_{pq} \cong \frac{2(ka)^3 (kb)^3}{15 p^2 q^2 \pi^5} \left\{ 1 - \frac{5k^2 ab}{64} \left[\left(1 - \frac{24}{(p\pi)^2}\right) \frac{a}{b} + \left(1 - \frac{24}{(q\pi)^2}\right) \frac{b}{a} \right] \right\}.$$
(2.86)

The odd-odd modes are the most effective radiators when ka and $kb \ll 1$. These expressions are used in Appendix B for a preliminary validation of the radiation efficiency of a benchmark structure.

2.7 Finite Element Method for Interior Problems

Consider a homogeneous, inviscid and compressible fluid occupying an acoustic region Ω_a bounded by a surface Γ. Such an interior problem is schematically shown in Fig. 2.7 in two dimensions. The bounding surface Γ is divided into non-overlapping segments such that $\Gamma = \Gamma_0 \cup \Gamma_V \cup \Gamma_Z$

2.7. FINITE ELEMENT METHOD FOR INTERIOR PROBLEMS

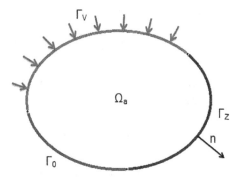

Figure 2.7. Interior coupled vibro-acoustic system

corresponding to an acoustically rigid boundary Γ_0, a vibrating boundary Γ_V with a harmonic normal velocity **v**, and a boundary Γ_Z covered with sound-absorbing material. Γ_Z is assumed to be locally reacting with a specific acoustic impedace Z. Such boundary conditions have been discussed in Section 2.5.

An approximate solution to the standard Helmholtz equation, given by Eq. (2.28), can be obtained by using the weighted residual technique [94]. If $\bar{p}$ is an approximate solution then

$$\nabla^2 \bar{p} + k^2 \bar{p} = \epsilon_1 \neq 0 \qquad (2.87)$$

where ϵ_1 is a non-zero error. The approximate solution $\bar{p}$ can be assumed to have the form

$$\bar{p} = \sum_{i=1}^{n} \phi_i(\mathbf{r}) q_i \qquad (2.88)$$

where $\phi_i(\mathbf{r})$ are prescribed functions of position **r** and q_i are unknown parameters to be determined. The shape functions do not fulfill the differential equation given by Eq. (2.28) and may violate the imposed boundary conditions. In a weighted residual formulation, the approximation errors are minimized in an integral sense. A solution is obtained by requiring the weighted average of this error to be zero.

Multiplying Eq. (2.87) by a weight $\bar{w}$ and integrating over the volume gives

$$\int_{\Omega_a} \bar{w} \nabla^2 \bar{p} d\Omega + \omega^2 \int_{\Omega_a} \frac{1}{c^2} \bar{w} \cdot \bar{p} d\Omega = 0. \qquad (2.89)$$

46 CHAPTER 2. FUNDAMENTALS IN STRUCTURAL ACOUSTICS

Introducing the Green's fist identity in the form

$$\int_{\Omega_a} \bar{w}\nabla^2 \bar{p}\, d\Omega = \int_{\Gamma} \mathbf{n}\cdot(\bar{w}\nabla\bar{p})\, d\Gamma - \int_{\Omega_a} \nabla\bar{w}\cdot\nabla\bar{p}\, d\Omega \qquad (2.90)$$

where $\mathbf{n}$ is the outward directed normal vector pointing out of the acoustic domain, Eq. (2.89) becomes

$$\int_{\Omega_a} \nabla\bar{w}\cdot\nabla\bar{p}\, d\Omega - \omega^2 \int_{\Omega_a} \frac{1}{c^2}\bar{w}\cdot\bar{p}\, d\Omega = \int_{\Gamma} \bar{w}\mathbf{n}\cdot\nabla\bar{p}\, d\Gamma. \qquad (2.91)$$

Since $\bar{p}$ is an approximate solution, from the boundary conditions given by Eqs. (2.30)–(2.32), it follows that

$$\mathbf{n}\cdot\nabla\bar{p} = \epsilon_2 \neq 0 \text{ on } \Gamma_0. \qquad (2.92)$$

$$\mathbf{n}\cdot\nabla\bar{p} + j\rho_0\omega\mathbf{n}\cdot\mathbf{v} = \epsilon_3 \neq 0 \text{ on } \Gamma_V. \qquad (2.93)$$

$$\mathbf{n}\cdot\nabla\bar{p} + j\rho_0\omega\frac{\bar{p}}{Z} = \epsilon_4 \neq 0 \text{ on } \Gamma_Z. \qquad (2.94)$$

Multiplying Eqs. (2.92)–(2.94) by the weight $\bar{w}$, integrating over the appropriate part of the surface, adding and equating the weighted average of each error to zero gives

$$\int_{\Gamma} \bar{w}\mathbf{n}\cdot\nabla\bar{p}\, d\Gamma = -j\omega\int_{\Gamma_V} \rho_0 \bar{w}\mathbf{n}\cdot\mathbf{v}\, d\Gamma - j\omega\int_{\Gamma_Z} \frac{\rho_0}{Z}\bar{w}\bar{p}\, d\Gamma. \qquad (2.95)$$

Substituting Eq. (2.95) into Eq. (2.91) gives

$$\int_{\Omega_a} \nabla\bar{w}\cdot\nabla\bar{p}\, d\Omega - \omega^2\int_{\Omega_a}\frac{1}{c^2}\bar{w}\cdot\bar{p}\, d\Omega + j\omega\int_{\Gamma_Z}\frac{\rho_0}{Z}\bar{w}\bar{p}\, d\Gamma = -j\omega\int_{\Gamma_V}\rho_0\bar{w}\mathbf{n}\cdot\mathbf{v}\, d\Gamma. \qquad (2.96)$$

Applying a Galerkin formulation, the weighting function is expanded in terms of the same set of shape functions used for the pressure expansion. Therefore, the weight $\bar{w}$ is taken to be of the form

$$\bar{w} = \sum_{i=1}^{n} \phi_i(\mathbf{r})w_i. \qquad (2.97)$$

Substituting Eqs. (2.88) and (2.97) into Eq. (2.96) results in a set of equations for the parameters q_i.

The most used approach for constructing the functions $\phi_i(\mathbf{r})$ is the finite element method. It approximates the acoustic field by a weighted sum of shape functions N_{a_i}. For each node i in the FE discretization there

2.7. FINITE ELEMENT METHOD FOR INTERIOR PROBLEMS 47

is an associate shape function N_{a_i} which has a non-zero value in each element domain to which node i belongs, while it is zero in all other element domains. The pressure distribution within each sub-volume is expressed by

$$p = \lfloor \mathbf{N}_a(x,y,z) \rfloor_e \{\mathbf{p}\}_e \qquad (2.98)$$

where $\lfloor \mathbf{N}_a(x,y,z) \rfloor_e$ is a row matrix of shape functions and $\{\mathbf{p}\}_e$ is a column matrix of nodal pressure values for element e. In addition, the weight function is also expressed in the form

$$w = \lfloor \mathbf{N}_a(x,y,z) \rfloor_e \{\mathbf{w}\}_e . \qquad (2.99)$$

For convenience, the overbar on the pressure p and weight w is excluded. The volume integrals in Eq. (2.96) are computed by the summation of integrations over individual elements. Substituting Eq. (2.98) and Eq. (2.99) gives

$$\int_{\Omega_e} \nabla \bar{w} \cdot \nabla \bar{p} \, d\Omega = \{\mathbf{w}\}_e^T [\mathbf{k}_a]_e \{\mathbf{p}\}_e \qquad (2.100)$$

where Ω_e is the volume of element e and

$$[\mathbf{k}_a]_e = \int_{\Omega_e} [\mathbf{B}]_e^T [\mathbf{B}]_e \, d\Omega \qquad (2.101)$$

with

$$[\mathbf{B}]_e = \nabla \lfloor \mathbf{N}_a \rfloor_e \qquad (2.102)$$

where ∇ denotes the gradient

$$\nabla = \left[\frac{\partial}{\partial x} \quad \frac{\partial}{\partial y} \quad \frac{\partial}{\partial z} \right]^T . \qquad (2.103)$$

We also define

$$\left(\frac{1}{c^2}\right) \int_{\Omega_e} wp \, d\Omega = \{\mathbf{w}\}_e^T [\mathbf{m}_a]_e \{\mathbf{p}\}_e \qquad (2.104)$$

where

$$[\mathbf{m}_a]_e = \left(\frac{1}{c^2}\right) \int_{\Omega_e} \lfloor \mathbf{N}_a \rfloor_e^T \lfloor \mathbf{N}_a \rfloor_e \, d\Omega. \qquad (2.105)$$

The variation of pressure over one surface of the element is approximated by $\lfloor \mathbf{N}_a(S) \rfloor_e$ obtained from the volume shape function $\lfloor \mathbf{N}_a \rfloor_e$. The contribution to the surface integral in Eq. (2.96) from each element in Γ_Z is thus given by

$$\int_{\Gamma_e} \left(\frac{\rho_0}{Z}\right) wp d\Gamma = \{\mathbf{w}\}_e^T [\mathbf{c}_a]_e \{\mathbf{p}\}_e \qquad (2.106)$$

where

$$[\mathbf{c}_a]_e = \int_{\Gamma_e} \left(\frac{\rho_0}{Z}\right) \lfloor \mathbf{N}_a(S) \rfloor_e^T \lfloor \mathbf{N}_a(S) \rfloor_e d\Gamma. \qquad (2.107)$$

The normal velocity distribution over an element of the surface Γ_V, assumed to be identical to one face of the acoustic element, is approximated by

$$\mathbf{n} \cdot \mathbf{v} = \lfloor \mathbf{N}_S \rfloor_e \{\mathbf{v}\}_e \qquad (2.108)$$

where $\{\mathbf{v}\}_e$ is a column matrix of normal velocities at the nodes. Therefore, the contribution to the surface integral in Eq. (2.96) from each element of the surface Γ_V is expressed by

$$\int_{\Gamma_e} \rho_0 w \mathbf{n} \cdot \mathbf{v} d\Gamma = \{\mathbf{w}\}_e^T [\mathbf{m}_{as}]_e \{\mathbf{v}\}_e \qquad (2.109)$$

where

$$[\mathbf{m}_{as}]_e = \int_{\Gamma_e} \rho_0 \lfloor \mathbf{N}_a(S) \rfloor_e^T \lfloor \mathbf{N}_S \rfloor_e d\Gamma. \qquad (2.110)$$

Summing the contribution from each individual element and substituting the results into Eq. (2.96) gives the following equation in terms of global matrices

$$\{\mathbf{w}\}^T \left[\mathbf{K}_a - \omega^2 \mathbf{M}_a + j\omega \mathbf{C}_a\right] \{\mathbf{p}\} = -j\omega \{\mathbf{w}\}^T [\mathbf{M}_{as}] \{\mathbf{v}\} \qquad (2.111)$$

where $\{\mathbf{p}\}$ is a column matrix of nodal pressure for the complete volume. Since $\{\mathbf{w}\}$ is arbitrary, then

$$\left[\mathbf{K}_a - \omega^2 \mathbf{M}_a + j\omega \mathbf{C}_a\right] \{\mathbf{p}\} = -j\omega [\mathbf{M}_{as}] \{\mathbf{v}\} \qquad (2.112)$$

where $\mathbf{K}_a$, $\mathbf{C}_a$ and $\mathbf{M}_a$ are the acoustic stiffness, damping and mass matrices. The solution of Eq. (2.112) yields the steady-state acoustic pressure values in the nodes of the FE discretization.

2.8 Coupled FE Formulation for Interior Vibro-Acoustic Systems

For an interior, fully coupled structural-acoustic case, the closed boundary surface Γ of the acoustic domain Ω_a includes a further elastic surface Γ_S causing the excitation of the acoustic medium. If the boundary surface Γ_S is the surface of a vibrating structure, then the resulting velocity distribution is given by the solution of the equations of motion of the structure, approximated by [58]

$$\left[\mathbf{K}_s - \omega^2 \mathbf{M}_s + j\omega \mathbf{C}_s\right] \{\mathbf{u}_s\} = \{\mathbf{f}\} \tag{2.113}$$

where $\{\mathbf{f}\}$ is a column matrix of equivalent nodal forces which accounts for mechanical forces $\{\mathbf{f}_m\}$ and the forces exerted by the acoustic medium on the structure $\{\mathbf{f}_a\}$. $\{\mathbf{u}_s\}$ is the structural displacement vector, $[\mathbf{M}_s]$, $[\mathbf{C}_s]$ and $[\mathbf{K}_s]$ are the structural mass, damping and stiffness matrices, respectively.

The force loading of the acoustic pressure on the structure can be considered as an additional normal load in the uncoupled structural model. This modifies the uncoupled structural Eq. (2.113) as follows

$$\left[\mathbf{K}_s - \omega^2 \mathbf{M}_s + j\omega \mathbf{C}_s\right] \{\mathbf{u}_s\} + [\mathbf{K}_{as}]\{\mathbf{p}\} = \{\mathbf{f}_m\} \tag{2.114}$$

where the acoustic nodal force vector is given by

$$\{\mathbf{f}_a\}_e = \int_{\Gamma_e} \lfloor \mathbf{N}_a(S) \rfloor_e^T \lfloor \mathbf{N}_a(S) \rfloor_e d\Gamma \{\mathbf{p}\}_e = \frac{1}{\rho_0} [\mathbf{m}_{as}]^T \{\mathbf{p}\}_e \tag{2.115}$$

and

$$\{\mathbf{f}_a\} = \frac{1}{\rho_0} [\mathbf{M}_{as}]^T \{\mathbf{p}\} = -[\mathbf{K}_{as}]\{\mathbf{p}\}. \tag{2.116}$$

The continuity of the normal structural velocity and the normal fluid velocity at the fluid–structure coupling interface can be regarded as an additional velocity input on the boundary surface Γ_S of the acoustic domain, as given by Eq. (2.112). The modified finite element model of the acoustic domain thus becomes

$$\left[\mathbf{K}_a - \omega^2 \mathbf{M}_a + j\omega \mathbf{C}_a\right] \{\mathbf{p}\} - \omega^2 [\mathbf{M}_{as}]\{\mathbf{u}_s\} = \{\mathbf{f}_q\} \tag{2.117}$$

where an acoustic point source of strength q_i located at node i is also considered within the volume. The source strength distribution Q_{vol} is given by

$$Q_{vol}(x,y,z) = j\omega\rho_0 q_i \delta(x_i, y_i, z_i) \tag{2.118}$$

where δ is the Dirac delta function at node i. The source vector $\{\mathbf{f}_q\}$ can thus be expressed as

$$\{\mathbf{f}_q\} = -j\omega\rho_0 \int_{\Omega_a} q_i [\mathbf{N}_a]^T \delta(x_i, y_i, z_i) d\Omega. \tag{2.119}$$

Provided that node i is not located on the boundary surface of Ω_a, all components of $\{\mathbf{f}_q\}$ are zero, except the component on row i which equals $j\omega\rho_0 q_i$.

Combining the acoustic FE model, given by Eq. (2.117), and the structural FE model, given by Eq. (2.114), yields the Eulerian FE/FE model for interior coupled vibro-acoustic systems

$$\left(\underbrace{\begin{bmatrix} \mathbf{K}_s & \mathbf{K}_{as} \\ 0 & \mathbf{K}_a \end{bmatrix}}_{\mathbf{K}_{fs}} + j\omega \underbrace{\begin{bmatrix} \mathbf{C}_s & 0 \\ 0 & \mathbf{C}_a \end{bmatrix}}_{\mathbf{C}_{fs}} - \omega^2 \underbrace{\begin{bmatrix} \mathbf{M}_s & 0 \\ \mathbf{M}_{as} & \mathbf{M}_a \end{bmatrix}}_{\mathbf{M}_{fs}} \right) \left\{ \begin{array}{c} \mathbf{u}_s \\ \mathbf{p} \end{array} \right\} = \underbrace{\left\{ \begin{array}{c} \mathbf{f}_m \\ \mathbf{f}_q \end{array} \right\}}_{\mathbf{f}_{fs}} \tag{2.120}$$

where $\mathbf{M}_s$ is the structural mass matrix, $\mathbf{M}_a$ is the acoustic mass matrix, $\mathbf{K}_s$ is the structural stiffness matrix, $\mathbf{K}_a$ is the acoustic stiffness matrix, $\mathbf{M}_{fs}$ is the coupling mass matrix, $\mathbf{K}_{fs}$ is the coupling stiffness matrix, $\mathbf{C}_s$ is the structural damping matrix, $\mathbf{C}_a$ is the acoustic damping matrix, $\mathbf{u}_S$ denotes the structural displacement, $\mathbf{p}$ denotes the nodal pressure in the fluid domain and $\mathbf{f}_b$ and $\mathbf{f}_a$ denote respectively the forces on the structural domain and the acoustic excitation. $\mathbf{M}_{sa}$, $\mathbf{K}_{sa}$ and $\mathbf{C}_{sa}$ are the fully coupled structural-acoustic matrices.

Equation (2.120), known as the Eulerian displacement-pressure (u/p) formulation, represents the most straightforward and compact form to address interior vibro-acoustic problems. The main drawback is that the coupled fluid-matrix system matrices are unsymmetric. This is due to the fact that the force loading of the fluid on the structure is proportional to the pressure while the force loading of the structure on the fluid is proportional to acceleration. Nevertheless, these coupled equations can be manipulated into a symmetric system of equations by taking the fluid velocity potential and the structural displacement as fundamental unknowns to describe the fluid and the structure domain respectively.

2.9 Summary

Sound radiation from vibrating panels plays a crucial role in the study of aircraft interior noise. When a structure is excited by an external mechanical or acoustical disturbance, its vibration transfers energy to the surrounding fluid medium. The radiated sound pressure level as well as the structural insertion loss and transmission loss are therefore attractive performance metrics for acoustic control of vibrating structures.

In this chapter, the fluid–structure coupling mechanism associated with the sound transmission through structures as well as the sound propagation in fluids have been presented. From the linear acoustic wave equation, the general acoustic radiation problem has been modelled by the Kirchoff–Helmholtz integral equation. For flat surfaces in a rigid baffle, this equation has been further simplified by the Rayleigh integral equation describing the radiated acoustic pressure field as a function of the structural vibration profile. Its derivation has been discussed by emphasizing the applicability and restrictions to real-time structural-acoustic control problems.

Sound radiation efficiency is an important issue to consider when studying sound transmission through multi-panel partitions. A radiation model has therefore been introduced to predict the sound power radiated from plate-like structures set in a rigid baffle. Two simplifying assumptions have been made in the model development: the surrounding medium has been assumed to be air and the radiated sound field has been evaluated in the far field. The former assumption is valid for uncoupled structural-acoustic problems. The result is that the acoustic pressure field is simply due to the uncoupled structural motion. The latter allows evaluating the total radiated sound power analytically from the characteristic impedance of the medium by integrating the acoustic intensity over a hemisphere. By substituting the eigenfunctions for an isotropic, homogeneous plate with simply supported boundary conditions, the radiated sound power has been analytically derived. A numerical method suitable for FE models has been also provided.

A short survey on the finite element formulation for interior coupled vibro-acoustic problems concludes the chapter. The resulting velocity distribution on the radiating surface can be associated with an equivalent set of elemental radiators considered as discrete monopole sources radiating into the free field.

Chapter 3

Transmission of Sound through Multiple Partitions

3.1 Introduction

Sound requires a medium to travel. The propagation of sound begins with the vibration of a fluid. When a propagating sound wave impinges a barrier, the incident wave is split into two components. Some of the wave is reflected back towards the source (reflected wave) and some passes through the barrier (transmitted wave). These basic terms are widely used in the literature to characterize the acoustic behaviour of structural barriers and acoustic enclosures.

Sound transmission loss of structures increases with the structural mass [15, 81]. By doubling the mass per unit area, sound transmission loss at a given frequency increases by 6 dB. In practice, however, the design of sound barrier systems is often driven by weight and space constraints. Thus, the use of multiple partitions can be more effective for those applications of sound transmission control that require a trade-off between light weight and high transmission loss performance over a broad frequency range.

Double- and triple-panel partitions provide an efficient solution for interior noise control due to the relatively low specific weight and high insertion loss. Even higher transmission loss values can be achieved by inserting acoustically absorptive material inside the fluid gap. However, due to the fluid-structure interaction, transmission loss of multi-wall structures suffers

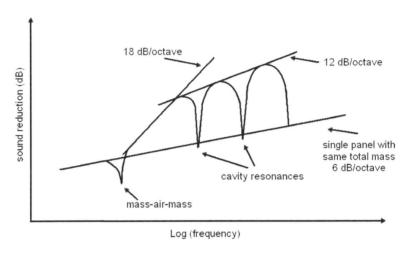

Figure 3.1. Transmission loss curve of a non-flexible double-panel partition

from the unfavourable effects of resonant noise transmission. At a certain frequency, known as the mass–air–mass resonance, the airspace, which acts like a spring between the partitions, dynamically couples the masses of the partitions, thus increasing the sound transmission in comparison to an equivalent single panel. Although this phenomenon occurs in a limited frequency range, it causes a significant drop in the sound transmission loss behaviour.

A typical sound transmission loss curve for infinite double panel partitions is plotted in Fig. 3.1. At very low frequencies, the double-panel partition behaves like a single panel with the same total mass. Around the mass–air–mass resonance frequency, the sound reduction is lower than that of a single panel, especially if no substantial mechanical damping is present. At higher frequencies the transmission loss rises at 18 dB per octave from the value it would have if simply controlled by the total mass of the partition. When the acoustic wavelength becomes of the same order of the dimension of the cavity between the panels, the curve increases with 12 dB per octave and shows dips at the cavity acoustic resonances.

In this chapter, the transmission of sound through double- and triple-panel partitions is investigated. A preliminary insight is given by assuming the panels are of infinite size. First, an engineering analysis of the mass–air–mass resonance phenomenon is provided by modelling the panel partitions as a system of masses (plates) and springs (air gaps). Next, the low-frequency behaviour of the coupled vibro-acoustic systems is studied

3.2. ANALYTICAL MODELLING

by introducing the modal coupling theory. A coupled structural-acoustic model of a flat triple panel is hence derived and an analytical model of sound transmission is developed in order to predict the sound reduction capabilities. Numerical evaluations of the sound transmission loss behaviour of finite plate-like multi-wall structures in idealized boundary conditions are also presented. They are based on a hybrid FEM/Rayleigh methodology describing the free field radiation from elemental planar monopole sources associated with the normal velocity distribution on the radiating surface. A summary concludes the chapter.

3.2 Analytical Modelling of Multi-Panel Partitions

Sound transmission of normal and oblique incident sound waves through infinite partitions is widely addressed in the literature [15, 81, 84]. *Vice versa*, there is a growing need for more reliable modelling and control concepts applied to multiple partitions due to the high practical importance in noise control applications. Sound transmission of plane sound waves through infinite-sized triple-panel partitions is studied in [95]. Impedance-based prediction models of low-frequency sound transmission through triple-panel partitions are also presented in [54].

In this section, the analysis of idealized infinite double-panel systems is introduced in order to describe the mechanism associated with the mass–air–mass resonance phenomenon. At such frequencies the two panels move out of phase, resulting in a significant reduction in the transmission loss behaviour. A brief description of the well-known mass law for single panels is also given. For more details the reader is referred to [81].

A general overview of the modal coupling theory applied to the fluid-structure interaction of triple wall structures is also presented. The steady state response is described by an infinite series of eigenfunctions. The partitions are supposed baffled in an infinite rigid wall and the coupled structural-acoustic response represented in terms of the summation of structural and acoustic mode shapes.

3.2.1 *Infinite panels*

A first insight into the behaviour of multi-panel partitions can be gained by considering a single panel of infinite size. Assume a uniform, unbounded, non-flexible partition of mass per unit area m mounted upon a viscously

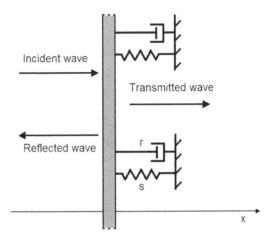

Figure 3.2. Idealized model of normal incidence transmission through a single panel partition

damped, elastic suspension having stiffness s and damping coefficient r per unit area. The idealized system is shown in Fig. 3.2. The partition separates two fluids with the same acoustic impedance $\rho_0 c$. A plane sound wave of frequency ω is normally incident upon the sending side of the partition from the region x < 0. The incident pressure can be written as

$$p_i(x,t) = \widetilde{P}_i e^{j(\omega t - k_1 x)} \tag{3.1}$$

where $k_1 = \omega/c$.

Fahy [81] calculates the transmission coefficient of normally incident plane waves as follows

$$\tau = \frac{4}{[(\omega m - s/\omega)/(\rho_0 c)]^2 + (\omega_0 m \eta/(\rho_0 c) + 2)^2} \tag{3.2}$$

where $\omega_0 = \sqrt{s/m}$ and η is the structural loss factor. The reader is referred to [81] for specifics.

Well above the natural frequency ($\omega >> \omega_0$), the Eq. (3.2) reduces to

$$\tau \cong \left(\frac{2\rho_0 c}{\omega m}\right)^2 \tag{3.3}$$

so that

$$\text{TL} = 20 \, \log\left(\frac{\omega m}{2\rho_0 c}\right). \tag{3.4}$$

This equation is known as the normal incidence mass law and indicates that the transmission loss increases with 20 dB/decade with frequency and by 6 dB per doubling of the mass of the panel.

3.2. ANALYTICAL MODELLING

If the acoustic waves impinges the wall with an angle θ with respect to the normal axis of the plate, then the transmission loss of a single panel of infinite size can be calculated by [81]

$$\text{TL} = 10 \, \log \left(1 + \frac{\omega^2 m^2 \cos^2 \theta}{4 \rho_0^2 c^2} \right). \tag{3.5}$$

For a fixed frequency below the coincidence, the transmission loss varies with the angle of incidence of the plane wave.

Mass–air–mass resonance frequencies

A set of simplified dynamic models can be developed to understand the sound transmission mechanism of normally incident plane waves through infinite double-panel or triple-panel partitions. In principle, at very low frequencies, a multi-wall panel behaves like a single partition with a total mass $M = \sum_{i=1}^{n} m_i$, where n is the number of partitions, and m_i is the mass per unit area of the ith partition. As the frequency increases, the coupling effect provided by the air chambers cannot be neglected, nor can the stiffness of the partitions.

Infinite double or triple partitions generally show higher sound transmission loss than single panels, except around certain frequencies, usually referred to as mass–air–mass resonance frequencies. At such frequencies, the panels oscillate in anti-phase against the stiffness of the acoustic medium in the cavities.

In order to explain this concept, consider a system of N degrees of freedom, constituted by N masses connected by k_{N+1} springs, as shown in Fig. 3.3.

The ith equation of motion of the system may be written as

$$\begin{array}{ll} m_i \ddot{x}_i + (k_i + k_{i+1})x_i - k_{i+1}x_{i+1} = 0 & i = 1 \\ m_i \ddot{x}_i - k_i x_{i-1} + (k_i + k_{i+1})x_i - k_{i+1}x_{i+1} = 0 & i = 2, ..., N-1 \, . \\ m_i \ddot{x}_i - k_i x_{i-1} + (k_i + k_{i+1})x_i = 0 & i = N \end{array} \tag{3.6}$$

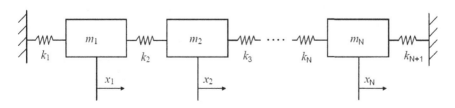

Figure 3.3. N degree of freedom system

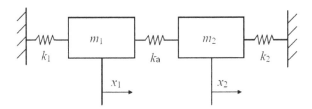

Figure 3.4. Idealized mass–spring–mass model for a double-panel partition

Similarly, well below the mass law region, a multi-panel partition can be modelled by a dynamic system constituted by m_i masses connected by springs with stiffness

$$k_{a_i} = \frac{\rho_0 c^2}{d_i} \qquad (3.7)$$

where d_i is the distance between two consecutive partitions.

Consider two uniform, non-flexible panels of mass per unit area m_1 and m_2, separated by a distance d and mounted upon elastic suspensions, having stiffness per unit area k_1 and k_2 respectively. The equations of motion of such a simplified mass–spring–mass model, shown in Fig. 3.4, may be written as:

$$\begin{aligned} m_1 \ddot{x}_1 + k_1 x_1 + k_a(x_1 - x_2) &= 0. \\ m_2 \ddot{x}_2 + k_2 x_2 + k_a(x_2 - x_1) &= 0. \end{aligned} \qquad (3.8)$$

The maximum transmission coefficient τ occurs at the low-frequency mass–air–mass resonance computed by neglecting the mechanical stiffness of the infinite-sized panels. Such a resonance frequency is expressed by:

$$f_0 = \frac{1}{2\pi} \sqrt{\frac{\rho_0 c^2}{d} \left(\frac{m_1 + m_2}{m_1 m_2} \right)} \qquad (3.9)$$

where d is the distance between the panels and m_1 and m_2 are the masses per unit area of the panels.

In the same way, the mass–air–mass–air–mass resonance frequencies of a triple-panel partition can be derived by considering a system of three masses, attached to each other by springs with stiffness k_{a_i}, with $i = 1, 2$. By solving the system in the frequency domain, it follows that

$$\begin{aligned}[m_1 m_2 m_3](\omega^2)^2 &- [k_{a_1} m_3(m_1 + m_2) + k_{a_2} m_1(m_2 + m_3)](\omega^2) \\ &+ \cdots + [k_{a_1} k_{a_2}(m_1 + m_2 + m_3)] = 0 \end{aligned} \qquad (3.10)$$

where $k_{a_1} = \rho_0 c^2 / d_1$ and $k_{a_2} = \rho_0 c^2 / d_2$ are the stiffness of the air chambers.

3.2. ANALYTICAL MODELLING

The resonance frequencies of the equivalent mass–spring–mass–spring–mass system are:

$$f_{1,2} = \frac{1}{2\pi} G \sqrt{A \pm \sqrt{A^2 - B}} \qquad (3.11)$$

where $G = \sqrt{\rho_0 c^2}$ and

$$A = \frac{1}{2m_2}\left(\frac{m_1 + m_2}{m_1 d_1} + \frac{m_2 + m_3}{m_3 d_2}\right); \; B = \frac{m_1 + m_2 + m_3}{m_1 m_2 m_3 d_1 d_2}. \qquad (3.12)$$

It should be noted that, at these frequencies, the pressure ratio of the transmitted to the incident sound wave becomes infinite. The validity of this assumption will be verified in Section 3.3. This condition is fulfilled if the denominator in Eq. (3.89) equals zero and the bending stiffness of the panels is neglected.

3.2.2 Modal coupling theory

As detailed in Section 2.8, vibrating panels coupled with an enclosed fluid domain can be analysed by using the finite element method (FEM). However, the unsymmetric, coupled Eulerian finite element (u/p) formulation can be computationally expensive and scarcely suitable for iterative structural and acoustic design modifications or for providing computationally efficient lower order models suitable for control applications via state-space techniques. Therefore, several techniques have been proposed to speed up the solution to a fully coupled finite element model [96, 97]. The modal coupling technique, described in the following, is one such method. It consists of using the uncoupled structural and acoustic modes of the system arising from the respective uncoupled eigenvalue problems to compute the coupled vibro-acoustic response. Although well suited for weakly coupled systems, this approach, originally proposed by Vacaitis [98] and Desmet et al. [99], is also used by Carneal et al. [100] and Kaiser et al. [101] for modelling the vibro-acoustic response of double panel systems.

The Panels

According to the Kirchoff plate theory, the motion of a thin rectangular plate at any time instant t is governed by the partial differential equation

$$D\nabla^4 w(\mathbf{r}_s, t) + \rho_s \frac{\partial^2 w(\mathbf{r}_s, t)}{\partial t^2} = f(\mathbf{r}_s, t) + p(\mathbf{r}_s, t) \qquad (3.13)$$

where $\mathbf{r}_s$ is an arbitrary location on the surface of the structure, $w(\mathbf{r}_s,t)$ is the displacement of the plate in the z-direction, $f(\mathbf{r}_s,t)$ is the distribution of mechanically applied force per unit area and $p(\mathbf{r}_s,t)$ is the pressure distribution on the surface.

The structural displacement can be expressed in terms of a summation over the *in vacuo* normal modes as

$$w(\mathbf{r}_s) = \sum_{p=1}^{\infty} W_p \psi_p(\mathbf{r}_s) \tag{3.14}$$

where ψ_p and W_p are the mode shape and the modal participation factor of the pth structural mode, respectively. Note that the time-dependent term $e^{j\omega t}$ has been omitted for brevity.

If simply supported boundary conditions are assumed for the rectangular plate of area $A = a \times b$, the mode shapes assume the well-known form

$$\psi_p(\mathbf{r}_s) = \psi_{mn}(\mathbf{r}_s) = \sin\left(\frac{m\pi x}{a}\right)\sin\left(\frac{n\pi y}{b}\right) \quad n,m = 1, 2, \ldots, \infty \tag{3.15}$$

where ψ_{mn} is the (m,n)th mode shape function which satisfies

$$D\nabla^4 \psi_{mn}(\mathbf{r}_s) - \omega_{mn}^2 \rho_s \psi_{mn}(\mathbf{r}_s) = 0. \tag{3.16}$$

The substitution into Eq. (3.13) of Eq. (3.14), together with the application of Eq. (3.16) followed by the multiplication by $\psi_{pq}(\mathbf{r}_s)$ and integration over the structural surface yields the modal equations of motion of the structure

$$\ddot{W}_{mn} + \omega_{mn}^2 W_{mn} = \frac{1}{M_{mn}}(F_{mn} + P_{mn}) \tag{3.17}$$

where F_{mn} and P_{mn} are the generalized force and pressure respectively, given by

$$F_{mn} = \int_A f(\mathbf{r}_s)\psi_{mn}(\mathbf{r}_s)\,dA \tag{3.18}$$

$$P_{mn} = \int_A p(\mathbf{r}_s)\psi_{mn}(\mathbf{r}_s)\,dA. \tag{3.19}$$

M_{mn}, usually referred to as the modal-generalized mass, is defined by

$$\int_A \rho_s \psi_{mn} \psi_{pq}\,dA = \begin{cases} M_{mn} & m=p,\ n=q \\ 0 & \text{otherwise} \end{cases}. \tag{3.20}$$

3.2. ANALYTICAL MODELLING

The cavity

The sound pressure p in a cavity is governed by the inhomogeneous wave equation

$$\nabla^2 p(\mathbf{r}, t) - \frac{1}{c^2} \frac{\partial^2 p(\mathbf{r}, t)}{\partial t^2} = -\rho_0 \frac{\partial q(\mathbf{r}, t)}{\partial t} \quad (3.21)$$

where q represents the distribution of source volume velocity per unit volume including effects of the boundary surface vibration.

The acoustic pressure in the acoustic cavity of volume V can be described in terms of a summation of the acoustic modes of the fluid volume as

$$p(\mathbf{r}) = \sum_{n=0}^{\infty} P_n \phi_n(\mathbf{r}) \quad (3.22)$$

where $\mathbf{r}$ is an arbitrary location within the volume and ϕ_n and P_n are the mode shape and the modal participation factor of the nth acoustic mode, respectively. For a rigid walled cavity, the mode shapes are given by

$$\phi_n(\mathbf{r}) = \phi_{ijk} = \cos\left(\frac{i\pi x}{l_x}\right) \cos\left(\frac{j\pi y}{l_y}\right) \cos\left(\frac{k\pi z}{l_z}\right) \quad i, j, k = 0, 1, 2, \ldots, \infty \quad (3.23)$$

where l_x, l_y and l_z are the cavity dimensions.

The modal equations for the response of the fluid can thus be written as

$$\ddot{P}_n + \omega_n^2 P_n = \frac{\rho_0 c^2}{M_n} F_n \quad (3.24)$$

where the modal normalization factor M_n is given by

$$\int_V \phi_m \phi_n \, dV = \begin{cases} M_n & m = n \\ 0 & \text{otherwise} \end{cases} \quad (3.25)$$

and the modal force is given by

$$F_n = \int_V \dot{q} \phi_n dV - \int_S \ddot{w} \phi_n dS \quad (3.26)$$

where the first term describes the sources in the cavity and q is the source strength with units of volume velocity. The second term is due to the flexible walls and $\ddot{w}$ is the normal acceleration of the boundaries. Both terms are projected on the nth cavity mode.

Vibro-acoustic coupling

In a coupled vibro-acoustic problem, the mechanical response of the panels and the acoustic response of the cavity are not independent. It is assumed that the modal expansion of the structural response, given by Eq. (3.14), is truncated to P structural modes and that of the acoustic response, given by Eq. (3.22), to N acoustic modes. Substituting Eq. (3.22) into Eq. (3.19) gives

$$P_p = \sum_{n=0}^{N} P_n \int_A \phi_n(\mathbf{r}_s)\psi_p(\mathbf{r}_s)\,dA = A \sum_{n=0}^{N} P_n C_{np} \qquad (3.27)$$

where the coupling factor C_{np} is a dimensionless coupling coefficient given by the integral of the product of the structural and acoustic mode shape functions over the surface of the structure. It is given by

$$C_{np} = \frac{1}{A} \int_A \phi_n(\mathbf{r}_s)\psi_p(\mathbf{r}_s)\,dA. \qquad (3.28)$$

The coupled modal equations of motion for the structure thus become

$$\ddot{W}_p + \omega_p^2 W_p = \frac{A}{M_p} \sum_{n=0}^{N} P_n C_{np} + \frac{F_p}{M_p}. \qquad (3.29)$$

Analogously, by substituting Eq. (3.14) into Eq. (3.26), the equations of the coupled response of the fluid are given by

$$\ddot{P}_n + \omega_n^2 P_n = -\frac{\rho_0 c^2 A}{M_n} \sum_{p=1}^{P} \ddot{W}_p C_{np} + \frac{\rho_0 c^2}{M_n} \dot{Q}_n. \qquad (3.30)$$

The equations for the fully coupled vibro-acoustic system can be formed into a matrix equation using Eqs. (3.28) to (3.30) as follows:

$$\begin{bmatrix} \mathbf{\Omega}_p & -A\mathbf{C} \\ -A\omega^2 \mathbf{C}^T & \mathbf{\Omega}_n \end{bmatrix} \begin{bmatrix} \mathbf{W} \\ \mathbf{P} \end{bmatrix} = \begin{bmatrix} \mathbf{F} \\ \dot{\mathbf{Q}} \end{bmatrix} \qquad (3.31)$$

where $\mathbf{\Omega}_p$ is a $(P \times P)$ diagonal matrix with elements $\Omega_{pp} = M_p(\omega_p^2 - \omega^2)$ and $\mathbf{\Omega}_n$ is a $(N \times N)$ diagonal matrix with elements $\Omega_{nn} = \frac{M_n}{\rho_0 c^2}(\omega_n^2 - \omega^2)$. The off diagonal elements represent the cross coupling between structure and fluid. $\mathbf{C}$ is a $(P \times N)$ matrix formed by the elements C_{np}. $\mathbf{W}$ and $\mathbf{P}$ are the modal amplitude vectors and $\mathbf{F}$ and $\dot{\mathbf{Q}}$ are the possible inputs to the system. Damping can be also added on a modal basis [81].

3.2. ANALYTICAL MODELLING

3.2.3 Structure-acoustic modal coupling in triple-panel partitions

In this section, the coupled structural-acoustic response of a triple-panel partition is described in terms of the structural and acoustic mode shapes. In this case, the modal coupling theory, detailed in Section 3.2.2, involves five subsystems, namely

- the dynamics of the incident panel
- the dynamics of the first cavity
- the dynamics of the middle panel
- the dynamics of the second cavity
- the dynamics of the radiating panel.

Each subsystem is relatively well known in the literature and it can be modelled individually. The coupled vibro-acoustic system can then be derived by properly assembling the individual subsystems.

The triple wall structure investigated as benchmark is shown in Fig. 3.5. It consists of three simply supported flexible panels and two rectangular fluid-filled cavities enclosed by four rigid side walls. From the side of sound incidence to the side of sound radiation the panels are referred to as

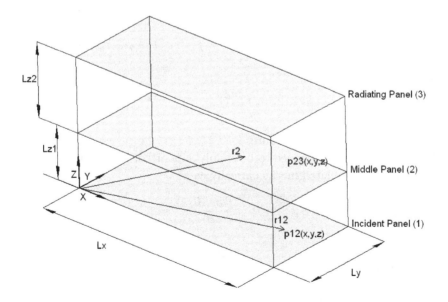

Figure 3.5. Triple panel partition

incident panel, *middle panel* and *radiating panel* respectively. The rectangular cavities have a volume $V_1 = L_x \times L_y \times L_{z_1}$ and $V_2 = L_x \times L_y \times L_{z_2}$. The sound field between the incident panel and the middle panel is defined as $p_{12}(\mathbf{r}_{12})$ while that between the middle panel and the radiating panel is referred to as $p_{23}(\mathbf{r}_{23})$. The pressures acting on the incident and the radiating panel are p_i and p_e respectively. The displacement distributions on the panels are indicated as $w_1(\mathbf{r}_{s_1})$ $w_2(\mathbf{r}_{s_2})$ and $w_3(\mathbf{r}_{s_3})$ respectively from the side of sound incidence to the side of sound radiation. The time dependence term $e^{j\omega t}$ is omitted for brevity.

Structural motion equations

According to the Kirchoff plate theory, the z-normal displacement w_1 at an arbitrary location $\mathbf{r}_{s_1}$ on the incident panel located in plane $z = 0$ is governed by the equation [81]

$$D_1 \nabla^4 w_1(\mathbf{r}_{s_1}, t) + \rho_{s_1} \frac{\partial^2 w_1(\mathbf{r}_{s_1}, t)}{\partial t^2} = p_i(\mathbf{r}_{s_1}, t) - p_{12}(\mathbf{r}_{s_1}, t) \qquad (3.32)$$

where w_1 is the transverse displacement, p_{12} is the cavity pressure loading the incident panel and p_i is the external pressure load.

In a similar way, the motion of the two panels, located in plane $z = L_{z1}$ and $z = L_{z2}$ is governed by the equations

$$D_2 \nabla^4 w_2(\mathbf{r}_{s_2}, t) + \rho_{s_2} \frac{\partial^2 w_2(\mathbf{r}_{s_2}, t)}{\partial t^2} = p_{12}(\mathbf{r}_{s_2}, t) - p_{23}(\mathbf{r}_{s_2}, t) \qquad (3.33)$$

and

$$D_3 \nabla^4 w_3(\mathbf{r}_{s_3}, t) + \rho_{s_3} \frac{\partial^2 w_3(\mathbf{r}_{s_3}, t)}{\partial t^2} = p_{23}(\mathbf{r}_{s_3}, t) - p_e(\mathbf{r}_{s_3}, t) \qquad (3.34)$$

where p_e is the external pressure.

If the panels are assumed to be simply supported, the boundary conditions can be formulated mathematically as

$$w_i(0, y) = 0, \quad w_i(L_x, y) = 0, \quad \frac{\partial^2 w_i(0, y)}{\partial x^2} = 0, \quad \frac{\partial^2 w_i(L_x, y)}{\partial x^2} = 0,$$
$$i = 1, 2, 3.$$

$$w_i(x, 0) = 0, \quad w_i(x, L_y) = 0, \quad \frac{\partial^2 w_i(x, 0)}{\partial y^2} = 0, \quad \frac{\partial^2 w_i(x, L_y)}{\partial y^2} = 0,$$
$$i = 1, 2, 3.$$
$$(3.35)$$

3.2. ANALYTICAL MODELLING

For simply supported plates, the normal *in vacuo* undamped modes are

$$\psi_m(\mathbf{r}_{s_i}) = \sin\left(\frac{p\pi x}{L_x}\right)\sin\left(\frac{q\pi y}{L_y}\right) \tag{3.36}$$

(p,q) being the modal indices of the mth mode and $\mathbf{r}_{s_i}$, for $i = 1, 2, 3$, the respective plate location vector. The corresponding modal angular frequencies are

$$\omega_m^{(i)} = \left(\frac{D_i}{\rho_{s_i}}\right)^{1/2}\left[\left(\frac{p\pi}{L_x}\right)^2 + \left(\frac{q\pi}{L_y}\right)^2\right] \quad i = 1, 2, 3. \tag{3.37}$$

The displacement $w_i(\mathbf{r}_{s_i})$ can be expanded as follows

$$w_i(\mathbf{r}_{s_i}) = \sum_{m=1}^{\infty} W_m^{(i)} \psi_m(\mathbf{r}_{s_i}) \quad i = 1, 2, 3. \tag{3.38}$$

The modal expansion can be substituted into Eqs. (3.32) to (3.34) and the orthogonality relations between the undamped modes used to give

$$\left(k_{m^2}^{(1)} - k^2\right)W_m^{(1)} = \frac{1}{M_m^{(1)}c^2}\left[\int_A p_i\psi_m dA - \int_A p_{12}\psi_m dA\right] \tag{3.39}$$

$$\left(k_{m^2}^{(2)} - k^2\right)W_m^{(2)} = \frac{1}{M_m^{(2)}c^2}\left[\int_A p_{12}\psi_m dA - \int_A p_{23}\psi_m dA\right] \tag{3.40}$$

$$\left(k_{m^2}^{(3)} - k^2\right)W_m^{(3)} = \frac{1}{M_m^{(3)}c^2}\left[\int_A p_{23}\psi_m dA - \int_A p_e\psi_m dA\right] \tag{3.41}$$

where the modal mass $M_m^{(i)}$ is defined as

$$M_m^{(i)} = \rho_{s_i}\int_A \psi_m\psi_m dA \quad i = 1, 2, 3 \tag{3.42}$$

and $m = 1, 2, 3, \ldots, \infty$ and $k_m^{(i)} = \omega_m^{(i)}/c$ for $i = 1, 2, 3$.

Acoustic subsystem equations

In a cavity with no acoustic sources, the acoustic pressure $p_{12}(\mathbf{r}_{12})$ at an arbitrary location $\mathbf{r}_{12}$ of the first cavity and $p_{23}(\mathbf{r}_{23})$ at an arbitrary location $\mathbf{r}_{23}$ of the second cavity obey the standard Helmholtz equation [81]

$$\nabla^2 p(\mathbf{r}_{12}, t) - \frac{1}{c^2}\frac{\partial^2 p(\mathbf{r}_{12}, t)}{\partial t^2} = 0 \tag{3.43}$$

and

$$\nabla^2 p(\mathbf{r}_{23}, t) - \frac{1}{c^2}\frac{\partial^2 p(\mathbf{r}_{23}, t)}{\partial t^2} = 0. \tag{3.44}$$

Along the four rigid side walls, the cavity pressure must satisfy the boundary conditions

$$\frac{\partial p_{12}(0,y,z)}{\partial x} = 0 \quad \frac{\partial p_{12}(L_x,y,z)}{\partial x} = 0 \quad \frac{\partial p_{12}(x,0,z)}{\partial y} = 0$$

$$\frac{\partial p_{12}(x,L_y,z)}{\partial y} = 0 \quad (0 \le z \le L_{z1})$$

$$\frac{\partial p_{23}(0,y,z)}{\partial x} = 0 \quad \frac{\partial p_{23}(L_x,y,z)}{\partial x} = 0 \quad \frac{\partial p_{23}(x,0,z)}{\partial y} = 0$$

$$\frac{\partial p_{23}(x,L_y,z)}{\partial y} = 0 \quad (L_{z1} \le z \le L_{z2}). \tag{3.45}$$

Along the fluid–structure interfaces or the so-called wetted surfaces, the cavity pressure must also satisfy the following conditions

$$\frac{\partial p_{12}(x,y,0)}{\partial z} = \rho_0 \omega^2 w_1(x,y) \quad \frac{\partial p_{12}(x,y,L_{z1})}{\partial z} = \rho_0 \omega^2 w_2(x,y)$$

$$\frac{\partial p_{23}(x,y,L_{z1})}{\partial z} = \rho_0 \omega^2 w_2(x,y) \quad \frac{\partial p_{23}(x,y,L_{z2})}{\partial z} = \rho_0 \omega^2 w_3(x,y). \tag{3.46}$$

The acoustic pressure $p_{12}(\mathbf{r}_{12})$ and $p_{23}(\mathbf{r}_{23})$ of the two fluid volumes with rigid boundaries is described in terms of the acoustic modes as

$$p_{12}(\mathbf{r}_{12}) = \sum_{n=0}^{\infty} P_n^{(12)} \phi_n^{(12)}(\mathbf{r}_{12}) \tag{3.47}$$

$$p_{23}(\mathbf{r}_{23}) = \sum_{n=0}^{\infty} P_n^{(23)} \phi_n^{(23)}(\mathbf{r}_{23}) \tag{3.48}$$

where $\mathbf{r}_{12}$ and $\mathbf{r}_{23}$ are the cavity location vectors ands

$$\phi_n^{(12)} = \cos\left(\frac{l\pi x}{L_x}\right)\cos\left(\frac{m\pi y}{L_y}\right)\cos\left(\frac{n\pi z}{L_{z1}}\right) \tag{3.49}$$

$$\phi_n^{(23)} = \cos\left(\frac{l\pi x}{L_x}\right)\cos\left(\frac{m\pi y}{L_y}\right)\cos\left(\frac{n\pi z}{L_{z2}}\right) \tag{3.50}$$

(l,m,n) being the modal indices of the nth acoustic mode. The corresponding modal angular frequency are

$$\omega_n^{(12)} = c\left[\left(\frac{l\pi}{L_x}\right)^2 + \left(\frac{m\pi}{L_y}\right)^2 + \left(\frac{n\pi}{L_{z1}}\right)^2\right]^{1/2} \tag{3.51}$$

3.2. ANALYTICAL MODELLING

and

$$\omega_n^{(23)} = c\left[\left(\frac{l\pi}{L_x}\right)^2 + \left(\frac{m\pi}{L_y}\right)^2 + \left(\frac{n\pi}{L_{z_2}}\right)^2\right]^{1/2}. \tag{3.52}$$

Applying the Green's second identity to Helmholtz equation yields

$$\int_V (p\nabla^2\phi_n - \phi_n\nabla^2 p)dV = \int_A \left(p\frac{\partial\phi_n}{\partial n} - \phi_n\frac{\partial p}{\partial n}\right)dA \tag{3.53}$$

n being the positive outward normal component.

For each cavity, Eq. (3.53) can be written as

$$\left(k^2 - k_n^{(12)2}\right)P_n^{(12)}\int_{V_1}\phi_n^{(12)}\phi_n^{(12)}dV$$

$$= \rho_0\omega^2\int_A \phi_n^{(12)}w_1 dA - \rho_0\omega^2\int_A \phi_n^{(12)}w_2 dA \tag{3.54}$$

$$\left(k^2 - k_n^{(23)2}\right)P_n^{(23)}\int_{V_2}\phi_n^{(23)}\phi_n^{(23)}dV$$

$$= \rho_0\omega^2\int_A \phi_n^{(23)}w_2 dA - \rho_0\omega^2\int_A \phi_n^{(23)}w_3 dA \tag{3.55}$$

for $n = 1, 2, \ldots, \infty$; $k_n^{(12)} = \omega_n^{(12)}/c$, $k_n^{(23)} = \omega_n^{(23)}/c$ are the wavenumber associated with the nth acoustic mode.

Due to the structural-acoustic coupling, Eqs. (3.39) to (3.41) include terms which describe the pressure load acting on the plates, while Eqs. (3.54) and (3.55) incorporate the plate displacements as acoustic source terms. The solution to the coupled system can be obtained by substituting the modal expansion of p_{12} and p_{23} and w_1, w_2, w_3 respectively.

From Eq. (3.38), the structural-acoustic coupling terms in Eqs. (3.54) and (3.55) become

$$\int_A \phi_n^{(12)}w_i dA = \int_A \phi_n^{(12)}\sum_{m=1}^{\infty}W_m^{(i)}\psi_m dA$$

$$= \sum_{m=1}^{\infty}W_m^{(i)}\int_A \psi_m\phi_n^{(12)}dA$$

$$= \sum_{m=1}^{\infty}W_m^{(i)}A\,C_{nm}^{(12)} \tag{3.56}$$

for $i = 1, 2$ and

$$\int_A \phi_n^{(23)} w_j \, dA = \int_A \phi_n^{(23)} \sum_{m=1}^{\infty} W_m^{(j)} \psi_m \, dA$$

$$= \sum_{m=1}^{\infty} W_m^{(j)} \int_A \psi_m \phi_n^{(23)} \, dA$$

$$= \sum_{m=1}^{\infty} W_m^{(j)} A \, C_{nm}^{(23)} \quad (3.57)$$

for $j = 2, 3$.

The C_{nm} term represents the modal spatial coupling between the nth cavity mode and the mth plate mode. These coupling coefficients are based on the geometrical similarity of the structural and the acoustic basis functions over the fluid–structure interface surface.

$$C_{nm}^{(12)} = \int_A \psi_m \phi_n^{(12)} \, dA \quad (3.58)$$

$$C_{nm}^{(23)} = \int_A \psi_m \phi_n^{(23)} \, dA. \quad (3.59)$$

Defining the acoustic modal mass

$$M_n^{(12)} = \rho_0 \int_{V_1} \phi_n^{(12)} \phi_n^{(12)} \, dV \quad (3.60)$$

and

$$M_n^{(23)} = \rho_0 \int_{V_2} \phi_n^{(23)} \phi_n^{(23)} \, dV. \quad (3.61)$$

Equations (3.54) and (3.55) may be written as

$$\left(k^2 - k_n^{(12)2}\right) P_n^{(12)} = \frac{k^2 \rho_0^2 c^2 A}{M_n^{(12)}} \left[\sum_{m=1}^{\infty} C_{nm}^{(12)} W_m^{(1)} - \sum_{m=1}^{\infty} C_{nm}^{(12)} W_m^{(2)}\right] \quad (3.62)$$

$$\left(k^2 - k_n^{(23)2}\right) P_n^{(23)} = \frac{k^2 \rho_0^2 c^2 A}{M_n^{(23)}} \left[\sum_{m=1}^{\infty} C_{nm}^{(23)} W_m^{(2)} - \sum_{m=1}^{\infty} C_{nm}^{(23)} W_m^{(3)}\right] \quad (3.63)$$

for $n = 1, 2, \ldots, \infty$.

3.3. SOUND TRANSMISSION

In the same way, using the modal expansion of p_{12} and p_{23} and introducing $C_{nm}^{(12)}$ and $C_{nm}^{(23)}$, Eqs. (3.39) to (3.41) may be written as

$$\left(k_m^{(1)2} - k^2\right) W_m^{(1)} = \frac{1}{M_m^{(1)} c^2} \left[Pi_m - A \sum_{n=0}^{\infty} C_{nm}^{(12)} P_n^{(12)} \right] \quad (3.64)$$

$$\left(k_m^{(2)2} - k^2\right) W_m^{(2)} = \frac{A}{M_m^{(2)} c^2} \left[\sum_{n=0}^{\infty} C_{nm}^{(12)} P_n^{(12)} - \sum_{n=0}^{\infty} C_{nm}^{(23)} P_n^{(23)} \right] \quad (3.65)$$

$$\left(k_m^{(3)2} - k^2\right) W_m^{(3)} = \frac{1}{M_m^{(3)} c^2} \left[A \sum_{n=0}^{\infty} C_{nm}^{(23)} P_n^{(23)} - Pe_m \right] \quad (3.66)$$

for $m = 1, 2, \ldots, \infty$, where the generalized acoustic modal excitation is given by:

$$Pi_m = \int_A p_i \psi_m dA \quad (3.67)$$

and the external modal excitation by:

$$Pe_m = \int_A p_e \psi_m dA. \quad (3.68)$$

The system can be assembled in matrix form as the one described in Section 3.2.2. If the fluid is air, it is possible to neglect the external pressure by solving the *in vacuo* case without the radiated pressure term. For the general case involving external mechanical forces and acoustic sources in the cavities, they will be included in the right-hand side as generalized mechanical and acoustic terms.

3.3 Sound Transmission Through Infinite Triple Partitions

A generic triple-panel partition is depicted in Fig. 3.6. The set-up consists of three infinite and homogeneous partitions separated by the distances d_1 and d_2, respectively. The panels separate a fluid on both sides as well as in the cavities with specific acoustic impedances $\rho_0 c$. The fluid behaves isoentropically.

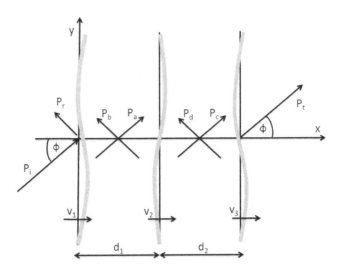

Figure 3.6. Transmission of oblique sound waves through an infinite triple wall partition

The incident, reflected and transmitted pressure sound field for the plane wave propagation can be written as

$$p_i(x,y) = P_i\, e^{-jk_x x}\, e^{-jk_y y} \qquad (3.69)$$
$$p_r(x,y) = P_r\, e^{jk_x x}\, e^{-jk_y y} = r\, P_i\, e^{jk_x x}\, e^{-jk_y y} \qquad (3.70)$$
$$p_t(x,y) = P_t\, e^{-jk_x x}\, e^{-jk_y y} \qquad (3.71)$$

where the term $e^{j\omega t}$ has been omitted for brevity. The reflection coefficient r is the ratio between the complex magnitude of the reflected wave to the complex magnitude of the incident wave. The wavenumber k is the magnitude of the wave vector **k** representing the direction of wave propagation given by the angle ϕ between **k** and the x axis. ϕ is determined by the relations

$$\sin\phi = k_y/k$$
$$\cos\phi = k_x/k.$$

In the first cavity, the pressure results from the wave transmitted by the first partition and from that reflected by the second partition:

$$p_a(x,y) = P_a\, e^{-jk_x x}\, e^{-jk_y y} \qquad (3.72)$$
$$p_b(x,y) = P_b\, e^{jk_x x}\, e^{-jk_y y}. \qquad (3.73)$$

Analogously, in the second cavity, the pressure results from the wave transmitted by the second partition and from the wave reflected by the

3.3. SOUND TRANSMISSION

third partition:
$$p_c(x,y) = P_c\, e^{-jk_x x}\, e^{-jk_y y} \tag{3.74}$$
$$p_d(x,y) = P_d\, e^{jk_x x}\, e^{-jk_y y}. \tag{3.75}$$

The plate velocities are given by the following relations
$$v_1(y) = V_1\, e^{-jk_y y} \tag{3.76}$$
$$v_2(y) = V_2\, e^{-jk_y y} \tag{3.77}$$
$$v_3(y) = V_3\, e^{-jk_y y}. \tag{3.78}$$

The boundary conditions for the particle velocity on the surface of the first partition follow as
$$v_1(0,y) = \frac{P_i(1-r)\cos\phi}{Z_0} e^{-jk_y y} \tag{3.79}$$
$$v_1(0,y) = \frac{(P_a - P_b)\cos\phi}{Z_0} e^{-jk_y y}. \tag{3.80}$$

For the second partition as
$$v_2(d_1,y) = \frac{\left(P_a\, e^{-jk_x d_1} - P_b\, e^{jk_x d_1}\right)\cos\phi}{Z_0} e^{-jk_y y} \tag{3.81}$$
$$v_2(d_1,y) = \frac{\left(P_c\, e^{-jk_x d_1} - P_d\, e^{jk_x d_1}\right)\cos\phi}{Z_0} e^{-jk_y y} \tag{3.82}$$

and for the third partition as
$$v_3(d_1+d_2,y) = \frac{\left(P_c\, e^{-jk_x(d_1+d_2)} - P_d\, e^{jk_x(d_1+d_2)}\right)\cos\phi}{Z_0} e^{-jk_y y} \tag{3.83}$$
$$v_3(d_1+d_2,y) = \frac{P_t \cos\phi\, e^{-jk_x(d_1+d_2)}}{Z_0} e^{-jk_y y}. \tag{3.84}$$

The bending wave equation for the plates in contact with the fluid yields
$$Z_{wp_1} V_1\, e^{-jk_y y} = \left(P_i(1+r) - (P_a + P_b)\right) e^{-jk_y y} \tag{3.85}$$
$$Z_{wp_2} V_2\, e^{-jk_y y} = \left(P_a e^{-jk_x d_1} + P_b e^{jk_x d_1} - \left(P_c e^{-jk_x d_1} + P_d e^{jk_x d_1}\right)\right) e^{-jk_y y} \tag{3.86}$$
$$Z_{wp_3} V_3\, e^{-jk_y y} = \left(P_c e^{-jk_x(d_1+d_2)} + P_d e^{jk_x(d_1+d_2)} - P_t e^{-jk_x(d_1+d_2)}\right) e^{-jk_y y} \tag{3.87}$$

where the structural wave impedance of the plates are

$$Z_{wp_i} = \frac{p}{j\omega w_i} = -j\left(D_i k^4 - m_i \omega^2\right)/\omega + D_i k^4 \eta_i/\omega \qquad (3.88)$$

for $i = 1, 2, 3$. w_i denotes the transversal displacement and D_i the bending stiffness of the ith partition of thickness h_i and density ρ_i, $m_i = \rho_i h_i$ the mass per unit area and $k = \omega/c_b$ with c_b the structural wave speed. Damping is represented by the introduction of a complex Young's modulus $\acute{E} = E(1 + j\eta)$, with η being defined as loss factor.

Resolving Eqs. (3.69) to (3.88), the pressure ratio of the transmitted to the incident sound waves reads

$$\frac{P_t}{P_i} = \frac{2\,Z_0{}^3\,(\cos(k\,(\mathrm{d}_1+\mathrm{d}_2)) + j\sin(k\,(\mathrm{d}_1+\mathrm{d}_2)))}{e^{jk\,(\mathrm{d}_1+\mathrm{d}_2)}\left(2\,Z_0{}^3 + Z_{wp_{123}} Z_0{}^2\,\cos(\phi)\right) + \tilde{Z}_{123}\,Z_0 \cos(\phi)^2 - \tilde{Z}_{wp_{123}}\sin(k\mathrm{d}_1)\sin(k\mathrm{d}_2)\cos(\phi)^3} \qquad (3.89)$$

with

$$\begin{aligned} Z_{wp_{123}} &= Z_{wp_1} + Z_{wp_2} + Z_{wp_3} \\ \tilde{Z}_{wp_{123}} &= Z_{wp_1} Z_{wp_2} Z_{wp_3} \\ \tilde{Z}_{123} &= \tilde{Z}_{13} + \tilde{Z}_{12} + \tilde{Z}_{23} \end{aligned} \qquad (3.90)$$

and

$$\begin{aligned} \tilde{Z}_{13} &= Z_{wp_1}\,Z_{wp_3}\,(j\sin(k(\mathrm{d}_1+\mathrm{d}_2))) \\ \tilde{Z}_{12} &= Z_{wp_1}\,Z_{wp_2}\,(-\sin(k\mathrm{d}_1)\sin(k\mathrm{d}_2) + j\cos(k\mathrm{d}_2)\sin(k\mathrm{d}_1)) \\ \tilde{Z}_{23} &= Z_{wp_2}\,Z_{wp_3}\,(-\sin(k\mathrm{d}_1)\sin(k\mathrm{d}_2) + j\cos(k\mathrm{d}_1)\sin(k\mathrm{d}_2)). \end{aligned} \qquad (3.91)$$

From Eq. (3.89) it is then possible to evaluate the sound transmission coefficient τ as

$$\tau(\omega) = \left|\frac{\tilde{P}_t}{\tilde{P}_i}\right|^2 \qquad (3.92)$$

and the sound transmission loss for an incident plane wave as

$$\mathrm{TL} = 10\log\left(\frac{1}{\tau(\omega)}\right). \qquad (3.93)$$

For a diffuse sound field excitation, plane waves from many directions excite the structural surface. The coefficient of diffuse sound transmission through the partition is defined by

$$\tau(\omega) = \int_0^{\phi_L} \tau_\phi(\omega)\sin(2\phi)d\phi \qquad (3.94)$$

where $\phi_L = 75° - 80°$. Diffuse field sound transmission loss is controlled by resonant and coincidence effects. Both lead to local transmission loss minima.

3.3. SOUND TRANSMISSION

3.3.1 Analysis of the derived equations

The general variation of normal and oblique incidence transmission of sound with frequency is here discussed by considering a triple wall partition of infinite extent consisting of two parallel glass panels (the incident and the middle panel respectively), one plexiglass panel (the radiating panel) and two enclosed air cavities. Such an example will be recalled in Chapter 5 for experimental validation. The resulting normal incidence sound transmission is compared with an existing impedance-based prediction model described in [95]. The numerical simulations are carried out in MATLAB®.

The triple-panel partition is asymmetric due to the difference in thickness: the two glass panels are 9.4 mm thick and the upper plexiglass panel is 6.4 mm thick. The acoustic cavities have also different depths. The masses per unit area are 7.61 kg/m^2 and 23.5 kg/m^2 for the plexiglass plate and the glass plates, respectively.

The sound transmission properties are also evaluated for three different double panel configurations: the 6.4 mm plexiglass panel and the 9.4 mm glass pane with an air gap of 16.6 mm, 36.2 mm and 52.8 mm, indicated as 6.4(16.6)9.4 (Double-glazed window I), 6.4(36.2)9.4 (Double-glazed window II) and 6.4(52.8)9.4 (Double-glazed window III) respectively, throughout the description of the results. The triple-glazed window is referred to as 6.4(36.2)9.4(16.6)9.4 (Triple-glazed window I). The normal incidence sound transmission through the double-panel partitions is investigated by using the prediction model developed by Fahy [81]. For such systems, the mass–air–mass resonance frequencies, computed by applying Eq. (3.9) and Eq. (3.11) are shown in Table 3.1.

Table 3.1 Analytical mass–air–mass resonances of the infinite and finite simply supported double- and triple-panel partitions

	Double-glazed window			Triple-glazed window	
	6.4(16.6)9.4	6.4(36.2)9.4	6.4(52.8)9.4	6.4(36.2)9.4(16.6)9.4	
				f_1	f_2
	(Hz)	(Hz)	(Hz)	(Hz)	(Hz)
Infinite panels	193.92	131.32	108.73	107.49	155.12
Finite panels	234.07	200.12	191.52	200.52	765.35

The sound transmission loss of normally incident plane waves through the infinite triple panel partition is shown in Fig. 3.7. Only the acoustic impedance of the two cavities and the inertia term of the panels are

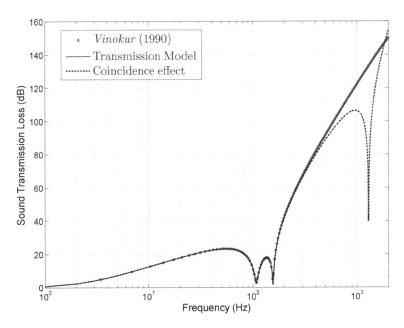

Figure 3.7. Transmission of normally incident plane waves through the unbounded triple panel partition

considered first. The partition exhibits extremely poor acoustical performance in the low frequency region due to the vibro-acoustic coupling between the infinite panels and the cavities. These low frequency dips may be ascribed to the theoretical mass–air–mass resonances, whose frequencies are below 200 Hz. Then, by considering the bending stiffness and the damping terms in Eq. (3.89), coincidence phenomena are included in the problem. At this specific frequency, expressed by Eq. (2.6), the transmission coefficient τ exhibits a local minimum. The greater the bending stiffness of the material and the less dense it is, the lower the coincidence dip. In the given example, the lowest coincidence occurs for the glass panel at a frequency of about 1.3 kHz. The structural properties are detailed in Section 3.4.2.

The influence of air cavity stiffness on the sound reduction index is also evaluated by varying the distances d_1 and d_2 between the panels. The achieved results are shown in Figs. 3.8 and 3.9, respectively. The transmission loss increases with the distance between the partitions while the mass–air–mass frequencies increase for decreasing values of partition separation distances.

3.3. SOUND TRANSMISSION

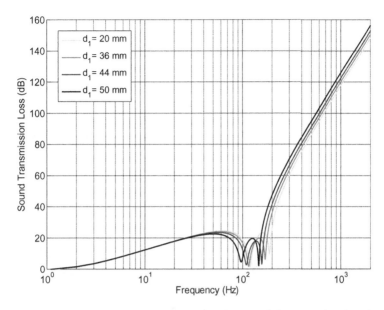

Figure 3.8. Sound transmission sensitivity to the distance d_1 between the plexiglass and the glass panel

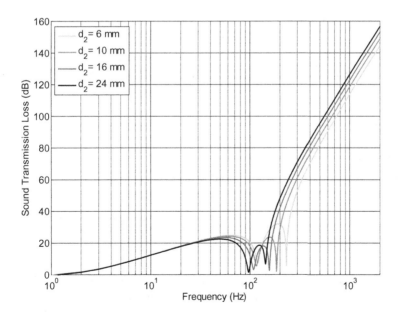

Figure 3.9. Sound transmission sensitivity to the distance d_2 between the two glass panels

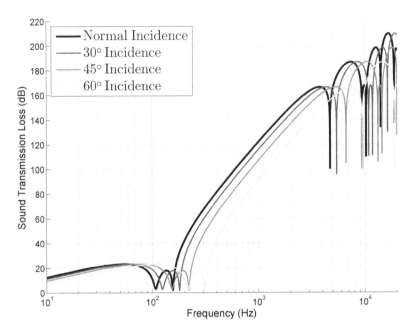

Figure 3.10. Transmission loss of an infinite triple panel partition for various angles of incidence

At higher frequencies, the sound transmission behaviour is governed by a succession of acoustic resonances occuring in the cavities. Resonance conditions arise for $kd\cos\phi = n\pi$ with n being any integer greater than zero. Figure 3.10 shows the influence of the angles of incidence (0°, 30°, 45°, 60°) on the sound transmission loss for frequencies up to 20 kHz. Transmission loss minima can be observed at each cavity resonance occuring at multiples of 4.7 kHz for normal incidence.

Finally, a comparison between the normal incidence sound transmission through the double- and triple-panel partitions is shown in Fig. 3.11. The corresponding mass–air–mass resonance frequencies are tabulated in Table 3.1. Below the mass–air–mass resonances, the double-leaf partitions behave like a single panel with mass which equals the sum of the masses m_1 and m_2. Similarly, the triple-wall structure behaves like a single panel with a mass m_t per unit area which equals the sum of the masses of the partitions. At low frequencies, the triple panel behaves like a double-glazed window with overall distance of both cavities. At frequencies above 200 Hz, the transmission loss properties significantly improve when switching from the single to the triple-panel partition.

3.3. SOUND TRANSMISSION

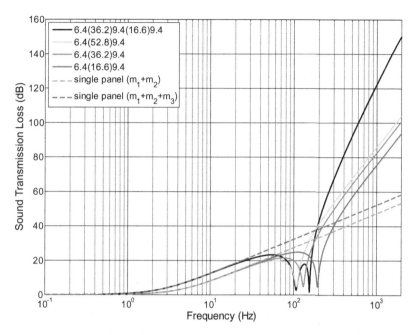

Figure 3.11. Normal incidence sound reduction index curves of infinite single, double- and triple-wall partitions

In the mass–air–mass frequency region, the transmission loss of a single panel is much higher than the sound reduction for an equivalent double- or triple-panel partition even though they have the same total amount of mass. On the other hand, the air cavities significantly improve the sound transmission loss of the multi-panel partitions at higher frequencies. At those frequencies, the transmission loss of the single panel increases with 20 dB per frequency decade (6 dB/octave), whereas the transmission loss of the double panel increases with 60 dB per frequency decade (18 dB/octave). These values are in agreement with the literature [81]. Results also show that the transmission loss of the triple panel partition rises asymptotically with about 110 dB per frequency decade (about 33 dB/octave).

Finite partitions

A preliminary estimate of the sound transmission properties of finite triple-wall structures can be obtained by assuming the mechanical stiffness of the panels is equal to $\omega_1 m_1^2$, $\omega_2 m_2^2$ and $\omega_3 m_3^2$, ω_1, ω_2 and ω_3 being the fundamental, *in vacuo* undamped resonance frequencies of the incident,

Table 3.2 Plexiglass panel properties and simulation data

Plexiglass Panel	
Young's modulus:	3.1 GPa
Poisson's ratio:	0.40
Density:	1190 kg/m^3
Dimensions:	0.20 × 0.32 cm
	6.4 mm thickness
Mesh:	640 ABAQUS® S4R shell elements
Boundary conditions:	Simply supported, rigid baffle

Table 3.3 Glass panel properties and simulation data

Glass Panel	
Young's modulus:	62 GPa
Poisson's ratio:	0.24
Density:	2480 kg/m^3
Dimensions:	0.20 × 0.32 cm
	9.4 mm thickness
Mesh:	640 ABAQUS® S4R shell elements
Boundary conditions:	Simply supported, rigid baffle

middle and radiating panel, respectively. This approach enables us to take into account the mechanical impedance of the panels along with the acoustic impedance $\rho_0 c$ and the acoustic stiffness $\frac{\rho_0 c^2}{d}$ of the cavities.

Following this approach, the sound transmission behaviour of the aforementioned triple- and the double-panel partitions has been investigated. The material properties of the plates are listed in Tables 3.2 and 3.3. The plates, whose dimensions are 200 × 320 mm, are assumed simply supported. The undamped resonance frequencies have been computed analytically. They are listed in Table 5.5.

The sound transmission loss curves for normally incident plane waves are shown in Fig. 3.12. Mechanical loss factors $\eta_1 = 0.03$ and $\eta_2 = \eta_3 = 0.0275$ have been used for the plexiglass and the glass panel respectively. Although the normal incidence is expected to overestimate the sound attenuation capabilities in comparison to a more classical diffuse incidence sound field, the results provide the necessary insight in the physical phenomenon of sound transmission through finite multi-wall structures. While at higher frequencies the transmission loss increases with a different slope depending on the mass law, the lower frequency performance is affected by resonance dips associated with the vibro-acoustic coupling. Active noise control techniques offer potential to reduce these low-frequency resonances, improving the transmission loss properties of the multi-wall structure. In order to

3.4. SOUND TRANSMISSION SIMULATION

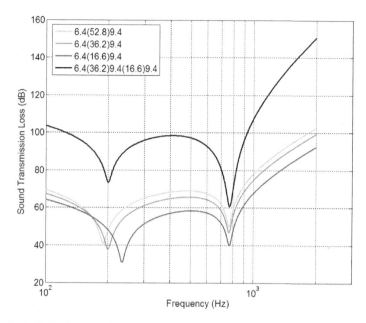

Figure 3.12. Preliminary evaluation of normal incidence sound transmission loss of finite double- and triple-panel partitions

identify the nature of these coupled modes, a coupled numerical analysis is provided in Section 3.4.2. The mass–air–mass resonance are also computed and listed in Table 3.1.

3.4 Sound Transmission Simulation

A numerical tool is proposed in this section to investigate the sound transmission loss of multiple panel partitions in the low frequency region. It is based on a four-step procedure, as follows:

- Simulation of diffuse sound field.
- Finite element modelling of the structural-acoustic coupling.
- Calculation of the acoustic radiation from the vibrating structure.
- Transmission loss evaluation.

The partition is assumed to be mounted in an infinite, rigid baffle separating two semi-infinite acoustic domains. As the acoustic medium on the incident and radiating side of the panels is air, a weak coupling is assumed with the exterior acoustic fluid. The acoustic radiation is thus separately evaluated from the interior vibro-acoustic response. The free field sound radiation is calculated by the Rayleigh integral. On the other hand, the fluid–structure coupling of the structural and acoustic subsystems is fully taken into account in the interior domain by means of a FEM-FEM approach. As illustrated in Fig. 3.11, due to the vibro-acoustic coupling, the resulting sound reduction index may be even lower than the one provided by a single panel with the same total mass.

The dynamic response of the vibro-acoustic system due to an acoustic diffuse excitation acting on the incident side of the multi-panel partition is computed first. Once the vibro-acoustic response on the radiating side is determined, the free field sound radiation is calculated by the Rayleigh integral. Finally, the transmission loss index is derived. The scheme adopted for the calculation of the transmission loss properties of multi-wall structures is shown in Fig. 3.13.

The sound power radiation and hence the sound reduction index can be calculated by implementing the elemental radiators theory detailed in Section 2.6.2. The normal surface velocity distribution in the radiating panel of

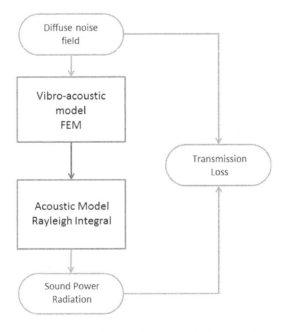

Figure 3.13. Sound transmission simulation procedure

3.4. SOUND TRANSMISSION SIMULATION

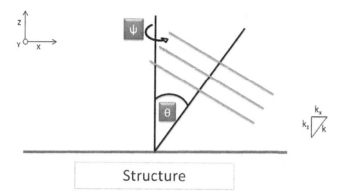

Figure 3.14. Acoustic plane waves incident on the surface of a structure

the multi-panel partition is used as input. The output is the free field sound radiation. In the following sections, some numerical applications involving double- and triple-panel partitions are detailed.

3.4.1 Diffuse acoustic excitation model

Diffuse fields are characteristic of reverberant spaces in which waves from many directions excite the surface of a structure. Reverberant chambers are constructed intentionally in acoustic test facilities for sound transmission loss measurements. In idealization, a diffuse sound field consists of an infinite number of uncorrelated plane progressive waves, with intensity uniformly distributed with respect to direction. It can be numerically simulated by assuming a large number, N, of plane waves having random angles of incidence, random magnitudes and random temporal phase angles. Figure 3.14 shows the nth plane wave, with a constant amplitude of P_n, incident on the surface of the structure at angles θ_n and ψ_n. The angles θ_n and ψ_n are random angles in spherical coordinates on the intervals $[0, \pi]$ and $[0, 2\pi]$ respectively.

The incident steady state pressure can be expressed in space and time as

$$p_n(x,y,z,t) = P_n \cos(\theta_n) e^{-jk_x x} e^{-jk_y y} e^{-jk_z z} e^{j(\omega t + \phi_n)} \quad (3.95)$$

where ϕ_n is distributed on the interval $[0, 2\pi]$ and

$$\begin{aligned} k_x &= k \sin(\theta_n) \cos(\psi_n) \\ k_y &= k \sin(\theta_n) \sin(\psi_n) \\ k_z &= k \cos(\theta_n). \end{aligned} \quad (3.96)$$

The random variables θ_n, ψ_n and P_n are unique for each of the N plane waves.

Assuming the pressure at the incident side as the one exerted by a rigidly blocked partition, the spatial pressure distribution exciting the structure can be written as

$$p_n(x, y, z, \omega) = 2P_n \cos(\theta_n) e^{j(\phi_n - k_x x - k_y y - k_z z)}. \tag{3.97}$$

In a finite element model, the total pressure acting on the surface of each element can be computed on the element centre of coordinates x_e, y_e and z_e

$$p_e(\omega) = \sum_{n=1}^{N} 2P_n \cos(\theta_n) e^{j(\phi_n + k_x x_e + k_y y_e + k_z z_e)} \tag{3.98}$$

where N is the number of plane waves used to approximate the diffuse field.

In order to compute the transmission loss, the ratio of the incident to transmitted sound power must be calculated. Starting from the expression of the incident pressure, the incident sound power can be computed from the incident intensity of each of the N plane waves.

The sound power incident on the surface of area A due to the nth plane wave is

$$\Pi_{i,n} = A \cos(\theta_n) \frac{[P_n \cos(\theta_n)]^2}{2\rho_0 c}. \tag{3.99}$$

The total incident sound power for all N plane waves is

$$\Pi_i = \sum_{n=1}^{N} \Pi_{i,n}. \tag{3.100}$$

With the radiated sound power computed separately by using the developed numerical code, the sound transmission loss can thus be calculated from the ratio of the incident and transmitted sound power:

$$TL = 10 \log_{10} \left(\frac{\Pi_i}{\Pi_t} \right). \tag{3.101}$$

3.4.2 Benchmark examples

In order to illustrate the applicability and suitability of the numerical hybrid FEM/Rayleigh methodology presented in Section 3.4, in what follows the proposed approach is applied to different numerical test cases in order to investigate the sound transmission loss behaviour of multi-wall structures.

Plexiglass single wall partition

The test case consists of a simply supported rectangular plate set in a rigid baffle. The model is a plexiglass plate having in-plane dimensions of 0.20 by 0.32 cm and a thickness of 6.4 mm. Material properties are illustrated in Table 3.2. The structural dynamic analysis is accomplished with the FEM-software ABAQUS®. In addition, a numerical tool is developed in MATLAB® for the post-processing of the results. The panel is modelled with 640 ABAQUS® S4 shell elements and it is simply supported on its edges.

A specific tool supplied by the commercial software ABAQUS® is employed to simulate the diffuse incident wave loading. According to [81], an approximated diffuse sound field can be generated by superimposing 250 temporally and spatially varying planar incident waves with incidence angles ranging from 0° to 78°. In the hypothesis of perfectly diffuse sound field, the incident sound power can be computed by Eq. (3.100). In ABAQUS,® a diffuse sound field is simulated by N^2 deterministically defined plane waves travelling in different directions according to N^2 discrete points arranged over the surface of an hemisphere, as shown in Fig. 3.15 [102]. This hemisphere is centered at a reference point, called standoff point. The incoming plane waves are oriented so that the incident power per solid angle approximates a diffuse incident field. The algorithm concentrates the points so that the incident waves in the diffuse field are normally incident, with fewer plane waves at oblique angles. The specified amplitude value and reference magnitude of the field are divided equally among the N^2 incident waves.

The structurally radiated sound power is computed by a specifically-developed MATLAB® code. Moreover, a numerical interface is developed in order to automatically process the output from the FE-solver. The ABAQUS® "*PU keyword" is used to collect the results in a ".dat" file with significant benefits in the whole computation time. A preliminary validation of the developed code is illustrated in Appendix B. In [103], the hybrid FEM/Rayleigh methodology is compared to commercially available BEM-tools, with results showing that it is sufficiently fast and well suited for iterative design processes. In [104], sound radiation from a baffled rectangular plate excited by a harmonic point force is compared with theory. ABAQUS® is used to simulate the structural response while the acoustic analysis is carried out by using LMS® Virtual.Lab and Sysnoise®.

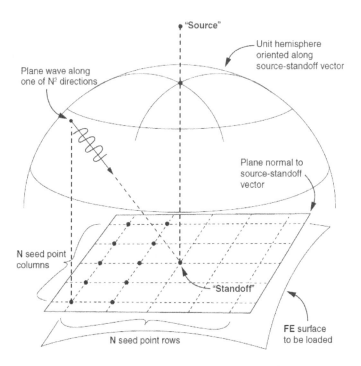

Figure 3.15. ABAQUS® Diffuse Loading model [102]

As detailed in Section 2.6, the sound power radiated from a rectangular plate set in an infinite rigid baffle can be expressed in terms of structural modes. For simple plate-like structures, these eigenfunctions can be determined analytically and used as a basis for a numerical comparison. A numerical diffuse-field transmission can be compared also with theoretical results.

Figure 3.16 shows a comparison between the analytical and numerical radiation efficiency of the investigated simply supported panel. Both analytical and numerical modes are used for the sound power calculation. The solution is based on a total of 133 structural modes in the range up to 12 kHz. A constant damping is used in the simulations. In addition, the structural radiation efficiency is calculated by loading the incident side of the panel with the diffuse sound field simulated in ABAQUS®. The resulting normal velocity distribution is used as input for calculating the sound power radiation. A good agreement is achieved between the curves. It should be noted that the modal average radiation efficiency exceeds unity well below the coincidence frequency occurring at 5.65 kHz.

3.4. SOUND TRANSMISSION SIMULATION

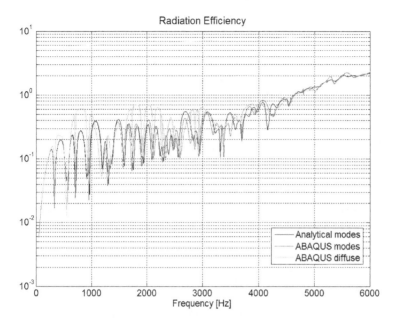

Figure 3.16. Comparison between the radiation efficiency computed for a simply supported panel

Figure 3.17 shows the transmission loss curves resulting from the three different methods investigated. The results show a reasonable agreement. The sound attenuation capability is low at the resonance frequencies and the minima depend on the structural damping. The dashed curves represent the mass law for 0° and 45° plane wave incidence. It can be noted that, well below the coincidence frequency, the calculated transmission loss curves converge towards the theoretical mass law resulting in a 6 dB increase per doubling of frequency.

Double- and triple-wall partitions

Finite double- and triple-wall panels are investigated to assess the sound transmission loss behaviour of multi-panel partitions. The sound attenuation of plane waves is numerically evaluated by taking into account the vibro-acoustic coupling between the panels and the enclosed cavity. The in-plane dimensions of the panels are 200 × 320 mm.

The triple-glazed window I and the double-glazed windows I, II and III are modelled in ABAQUS® by uniform S4 shell elements and simply supported on the edges. The fluid is modelled with linear hexahedral solid

86   CHAPTER 3. TRANSMISSION OF SOUND

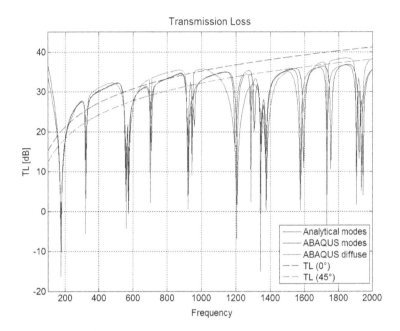

Figure 3.17. Transmission loss behaviour of a simply supported panel

Table 3.4 Fluid properties

Fluid	
Density:	1.225 kg/m^3
Bulk modulus:	0.142 MPa
Speed of sound:	340 m/s
Mesh:	3200 ABAQUS® Hexa AC3D8 elements
Boundary conditions:	Structural velocity or $v_n = 0$

elements of type AC3D8 matching the structural mesh of the panel. Materials and fluid properties are summarized in Table 3.2–3.4. The finite element model of the triple wall structure is shown in Fig. 3.18. The structure and fluid domain are coupled on the wetted surface by the normal components of velocity, whereas the acoustic rigid condition $\frac{\partial p}{\partial n} = 0$ is applied to all the surfaces of fluid not in contact with the structure. The vibro-acoustic dynamic response of the coupled structures is determined by a modal frequency analysis. A step size of 0.5 Hz is selected for the frequency resolution. Constant structural damping is assumed.

3.4. SOUND TRANSMISSION SIMULATION

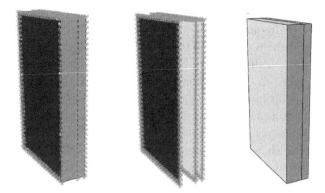

Figure 3.18. FE model of the simply supported triple wall structure: Assembly (left), panels (centre) and cavities (right)

Table 3.5 Numerical vibro-acoustic modes of the double- and triple-wall panels

	Double-glazed		Triple-glazed	
ID	6.4(16.6)9.4 (Hz)	6.4(36.2)9.4 (Hz)	6.4(52.8)9.4 (Hz)	6.4(36.2)9.4(16.6)9.4 (Hz)
1	221.52	198.55	191.79	198.66
2	309.87	317.01	318.98	317.20
3	—	—	—	529.67
4	551.88	541.26	538.27	541.68
5	548.41	554.32	555.84	554.33
6	568.87	570.53	570.87	570.56
7	692.55	697.97	699.31	697.97
8	753.93	752.08	751.53	750.78
9	—	—	—	757.20

Eigenvalue analysis

A modal analysis is performed in ABAQUS®. The resulting eigenfrequencies are listed in Table 3.5. In contrast to the results obtained for infinite panels, it should be noted that the finite triple-glazed window behaves more like a 6.4(36.2)9.4 than a 6.4(52.8)9.4 double-glazed panel. Due to the asymmetric configuration, with one rather small (16.6 mm) and one rather large (36.2 mm) pane distance, and due to the higher stiffness of the glass panes compared to the plexiglass one, the triple-panel acts as a double-panel with one pane as thick as the two near glass panes together. The influence of the outer glass pane is negligible, as confirmed by the close agreement between the coupled eigenfrequencies resulting from the 6.4(36.2)9.4 (Double-glazed window II) and the 6.4(36.2)9.4(16.6)9.4 (Triple-glazed window I) configurations, shown in Table 3.5. This behaviour is also in agreement with the preliminary mass–air–mass resonance estimations given in Table 3.1. Nevertheless, the additional mass of the 9.4 thick glass panel ensures significant benefits in the sound transmission loss index, as illustrated in Figure 3.12.

The mode shapes of the triple wall panel are shown in Fig. 3.19. The uncoupled mode shape of the plexiglass panel, from which the first coupled resonance originates, is the dominant mode at 198.66 Hz. At that frequency, the vibration of the middle glass panel is not of the same order. The acoustic cavities of the triple partition have the same (0,1,0) mode at 531.25 Hz, as listed in Table 3.6. Such an uncoupled cavity mode shape, from which both the coupled mode at 529.67 Hz and at 541.68 Hz originate, is clearly recognizable in the pressure distribution at those frequencies, as shown in Fig. 3.19. Mode no. 3 at 529.67 Hz is the combination of the (0,1,0) cavity mode of the (16.6) air gap and the uncoupled plate modes (1,2) of the two glass panes, with the plates moving out of phase. This means that for two corresponding points on both panels, the one moves upwards while the other moves downwards and *vice versa*. Mode no. 4 at 541.68 Hz is the combination of the (0,1,0) cavity mode of the (36.2) air gap and the uncoupled plate mode (1,2) of the plexiglass panel. The (2,1), (1,3) and (2,2) uncoupled plexiglass mode shapes dominate the vibration shapes of the modes no. 5, 6 and 7 respectively. Finally, the coupled modes at 750.78 Hz and 757.20 Hz are dominated by the (1,1) uncoupled plate modes of the two glass panes, which move in phase and in opposite phase respectively, as shown in Fig. 3.19. This behaviour is driven by the cavity stiffness in the (16.6) air gap and it corresponds to the second mass–air–mass resonance frequency of the triple-panel partition.

Table 3.6 lists the theoretical resonance frequencies of the acoustic cavities, computed by using Eq. (3.51), for the speed of sound of 340 m/s. They can be divided into acoustic cavity resonances with standing waves parallel to the panels and, at much higher frequencies, those with standing waves perpendicular to the panels. The latter provide higher sound transmission attenuation. The distance d between the partitions has no effect on the eigenfrequencies of the $(p, q, 0)$ modes, i.e. for all modes with wavenumber 0 in z-direction. The mode $(0, 0, 1)$ is the first mode for which the distance between the panels matters. This uncoupled mode is also calculated for the three different separation distances.

3.4. SOUND TRANSMISSION SIMULATION

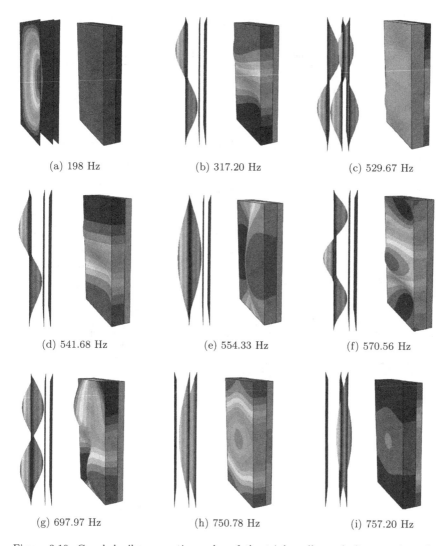

(a) 198 Hz (b) 317.20 Hz (c) 529.67 Hz

(d) 541.68 Hz (e) 554.33 Hz (f) 570.56 Hz

(g) 697.97 Hz (h) 750.78 Hz (i) 757.20 Hz

Figure 3.19. Coupled vibro-acoustic modes of the triple-wall panel: Structural mode shapes (left), pressure distribution within the cavities (right)

Transmission loss predictions

Numerical sound transmission loss simulations involving the idealized simply supported multi-panel systems are presented in this section. The incident sound power is obtained from a numerically simulated diffuse sound field excitation acting on the incident panel (glass panel). The resulting free

Table 3.6 Uncoupled acoustic cavity resonances

Modal indices	Rigid-walled acoustic frequency (Hz)	Air space (cm)
(0,0,0)	0	
(0,1,0)	531.25	
(1,0,0)	850	
(1,1,0)	1002.36	
(0,2,0)	1062.5	
(1,2,0)	1360.66	
(2,0,0)	1700	
(0,0,1)	3219.69	5.28
(0,0,1)	4696.13	3.62
(0,0,1)	10240.96	1.66

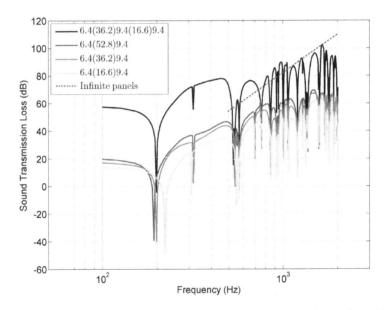

Figure 3.20. Diffuse field sound transmission loss of double- and triple-panel partitions

field sound radiation is computed from the surface velocity distribution achieved on the radiating panel (plexiglass panel). Results are shown in Fig. 3.20. In such simplified models, the only means of connection between the panels is through the coupling due to the air spaces. As a result, a number of local minima corresponding to the vibro-acoustic resonances, where the panel response is coupled with the fluid domain, are obtained in the whole frequency range. Such frequencies are in agreement with those listed in Table 3.5. The structural energy dissipation and the cavity absorption controlling the sound power radiation in the simulation model are intentionally kept at a minimum value in order to highlight the resonant contribution to the structural transmission loss which is intended to be actively controlled. By increasing the modal damping in the panels and in the acoustic cavities, the resonant components in the sound transmission can be reduced and the distinct minima in transmission loss increased. Nevertheless, the transmission loss contribution governed by the non-resonant mass law is not affected. The effect of increasing values of the partition separation distance is also evaluated. Compared to the transmission loss of Double-glazed window I, the transmission loss of the double-glazed window III is about 10 dB higher at frequencies above the mass–air–mass resonance. This is in agreement with the theoretical values found in the literature [81].

In order to analyze the effects of the panel boundedness on the sound transmission loss, a comparison is also made with the theoretical diffuse incidence sound reduction index obtained for an infinite triple-panel partition. The main differences are due to the large dips caused by the standing waves and the large responses occurring at the coupled structural-acoustic resonances.

3.5 Summary

The need for lightweight structures with enhanced transmission loss capabilities motivates theoretical and experimental efforts in designing multiple panel partitions separated by fluid gaps. Nevertheless, the vibro-acoustic behaviour of double- and triple-wall structures is strongly influenced by the fluid–structure coupling between the panels and the enclosed cavities.

In this chapter, a theoretical analysis of the fluid–structure interaction of double- and triple-panel partitions has been presented in order to provide an useful insight into the underlying mechanism of the vibro-acoustic coupling to be controlled by active control methods. As a first step, double- and triple-panel partitions have been modelled as a system of masses (plates) and springs (air gaps). The mass–air–mass resonance frequencies have been derived by neglecting the mechanical stiffness of infinite-sized panels. Next, the analysis has been extended to flexible plates separated by air gaps with idealized boundary conditions. An approach to model the structural-acoustic modal coupling in a triple-panel partition has been presented.

The analysis has been based on the modal expansion method combining the *in vacuo* structural modal response of the panels with the hard-walled acoustic modal response of the cavities.

An analytical model of sound transmission of normally and obliquely incident plane waves through unbounded triple-panel partitions has been developed and validated. With respect to the sound transmission models available in the literature, the developed model allows us to predict coincidence phenomena by modelling the structural impedance.

The description of a FEM-based simulation strategy to model the structure-acoustic behaviour of plate-like structures concludes the chapter. In this numerical procedure, a weak coupling is assumed between the structural and exterior acoustic fluid while the structural and acoustic responses, driving the sound transmission through the partitions, are fully coupled in the interior domain. The vibro-acoustic response of the structure is numerically modelled first. The resulting surface velocity distribution is then used as input for calculating the free field sound radiation with the Rayleigh integral. The result is an efficient tool capable of predicting the sound transmission loss properties of baffled structures.

The performance of the proposed sound transmission simulation strategy has been validated on simple test cases consisting of simply supported single and multiple partitions. The diffuse excitation has been modelled with a commercial software. The achieved analytical and numerical results have been in good agreement.

In the following chapters, the developed sound transmission loss prediction tool will be recalled to assess the augmented sound transmission loss capabilities of aircraft-type windows by active control.

Chapter 4

Adaptive Control of Sound Radiation

4.1 Introduction

Active control of sound was one of the earliest applications of electronics to the control of physical systems. First attempts are dated back to the early 20th century and were made by Paul Lueg in 1936 [67]. His arrangement was the first example of active noise cancellation of sound in a duct by utilizing a reference signal, captured and passed through a controller to drive a loudspeaker creating a counteracting acoustic field. This technique is now termed feedforward active noise control.

In 1953, Olson and May performed the first experiments on feedback active noise control [68]. Feedback systems are distinguished from feedforward systems in that the need for a reference signal correlated to the primary disturbance being controlled is eliminated. The error signal is directly fed back to the controller in order to create a zone of quiet around the error microphones [101, 105]. The main drawback associated with this type of control is the limited stability robustness. The general way to decrease the risk of instability is to use collocated feedback control, based upon actuator and sensor pairs [106]. Alternatively, in a model-based feedback control, the control algorithm can be reformulated as a feedforward problem by using an internal estimate of the primary plant [107].

A single-channel feedforward active control architecture is schematically illustrated in Fig. 4.1. Time-advanced information is available from the reference sensor. The primary disturbance is captured by the error sensor

94 CHAPTER 4. ADAPTIVE CONTROL OF SOUND RADIATION

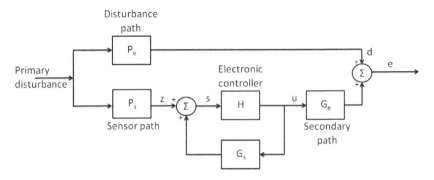

Figure 4.1. Block diagram of a single-channel feedforward active control system

via the disturbance path P_e to give d, and it is measured by the reference sensor via the sensor path P_s to give z. The output of the secondary source affects the error sensor via the secondary path G_e as well as the reference sensor via the feedback path G_s. The objective is to design the feedforward controller H such that the output from the error sensor e is minimized.

The advent of integrated circuits and the rapid development of cost-effective digital signal processing (DSP) chips have enabled digital active control technology to emerge as a practical alternative to passive treatments in actual aircraft applications. Active control solutions are potentially most effective at low frequencies where conventional, passive, noise control means cannot generally be applied without weight and space penalties.

A considerable amount of work on active noise control has been devoted to the development of methodologies for the design and realization of ANC systems in single-input-single-output (SISO) applications. Nevertheless, multichannel systems, based on multi-input-multi-output (MIMO) controllers, have been considered in various noise and vibration control applications involving both periodic and broadband disturbances [108]. However, memory requirements and computation loads as well as the mutual interaction between the secondary sources and both the detectors and observation sensors have limited the performance of such control architectures on large systems.

A multi-channel active noise control system for broadband noise disturbances has been tested on a light jet aircraft to enhance cabin comfort [109]. In [110], a controller design method to optimize the loudspeaker and microphone positions in a narrow-band system is presented to achieve a global noise reduction inside an aircraft passenger cabin. In order to create

4.1. INTRODUCTION

a zone of quiet around the passenger's head, relatively large loudspeaker arrangements are used. In [111, 112], active noise control is applied to reduce interior sound field in a propeller-driven military transport aircraft. As discussed in Section 1.2, this approach has been termed active noise control (ANC).

Similarly, many attempts have been made in the past at devising methods for tackling sound radiation from vibrating structures. In doing so, the generated sound is controlled by actuators driving the structure being controlled, sensors detecting the noise disturbance and a controller aiming at reducing the residual acoustic field. Such systems are mainly focused on controlling the most acoustically efficient radiating modes by using point force inputs directly attached to the structure rather than imposing a secondary, loudspeaker-generated sound field. Active structural-acoustic control is then realized either by modal suppression, meaning that during control all the modal amplitudes are reduced, or by modal restructuring leading to a residual structural response with lower radiation efficiency. In [113], active control of plate structures uses surface sensors polyvinylidene fluoride (PVDF) to estimate the acoustic radiation and piezoelectric elements as control actuators.

Due to the highly correlated reference signal, which is accessible by monitoring the engines, active noise reduction in propeller aircraft has been a typical subject of feedforward control. With the advances in adaptive structures, this approach has been extended to ASAC applications by providing the structure with integrated sensing and actuation capabilities. For jet aircraft, instead, reliable control strategies have yet to be demonstrated since many complex sources such as boundary layer noise and jet noise are involved in creating the disturbance in the cabin.

A feedforward algorithm relies on the knowledge of the transfer functions from the secondary sources to the error microphones (the so-called secondary paths or control paths) to generate the optimal control signals. For broadband excitations, accurate experimental models predicting the acoustic response at the error sensors are indispensable to enhancing the system performance [114]. Furthermore, these secondary paths can be subjected to changes during operation due to time-varying flight conditions or temperature variations. As a result, the control algorithm must provide sufficient robustness properties. An alternative way to face this problem is to measure the transfer functions on a regular basis during operation and adapt the controller accordingly. Such an approach, referred to as on-line identification, requires significant computational efforts on the control

96　CHAPTER 4. ADAPTIVE CONTROL OF SOUND RADIATION

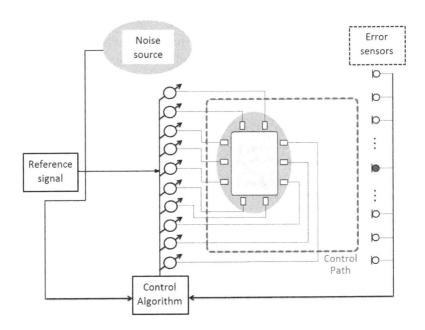

Figure 4.2. A multi-input-multi-output system for active control of sound radiation

system. A more comprehensive insight into adaptive signal processing and control is covered in [115].

This chapter starts with a brief overview of structural-acoustic control. After that, theory is applied to a benchmark application for local active sound transmission control. A schematic of the actuation and sensing approach is shown in Fig. 4.2. An adaptive feedforward multi-channel control is considered to reduce the radiated sound field at the error sensors. The adaptive control system consists of two main parts: a number of digital filters which drive the secondary sources and an adaptive controller which updates the filter coefficients in order to track variations in the noise radiation characteristics. The reference signal is taken directly from the primary source. Furthermore, there is no feedback from the secondary sources to the reference signal. This approach is particularly useful for periodic signals, like the propeller-induced noise in a turboprop aircraft cabin. The X-filtered version of the adaptive least mean squares (LMS) algorithm is considered and up to eight piezoelectric actuators and four microphones are utilized as control inputs and error sensors, respectively. An accurate experimental identification of the impulse response functions of the primary signals and the control actuators is performed over the entire frequency band of interest. In addition, as an alternative to the multi-channel version of the filtered-X LMS algorithm, the least maximum mean squares (LMMS) and the scanning error LMS algorithms are implemented and tested against different types of primary sound fields [116]. From harmonic disturbances to band-limited white noise excitations are considered. A variety of numerical cases are investigated by varying the number of control actuators and evaluating the performance in suppressing the noise transmitted through the smart system test-bed.

4.2 Control of Sound Radiation by Structural Actuators

Active structural control of sound radiation from vibrating structures requires a different strategy than pure active vibration control. The residual error signal that is minimized by the control system is an acoustical quantity and active vibration actuators are located close to the original vibration source(s) or on the vibration transmission path. When a radiating panel is excited by a primary force distribution and M secondary point forces, $f_{s1}, ..., f_{sM}$, then the complex amplitude of the nth mode at a single frequency can be written as [93]

$$a_n = a_{np} + \sum_{m=1}^{M} B_{nm} f_{sm} \qquad (4.1)$$

where B_{nm} is the coupling coefficient between the mth actuator and nth mode, which is proportional to the mode shape $\psi_n(x, y)$ at the point of application of the mth secondary force, and a_{np} is the amplitude of the nth mode due to the primary force.

Let us assume that the response of the radiating panel is sufficiently represented by the sum over a finite number (N) of modes. The complex mode amplitude vector of the N modes, $\mathbf{a}$, can then be written in vector form as

$$\mathbf{a} = \mathbf{a}_p + \mathbf{B}\mathbf{f}_s \qquad (4.2)$$

where $\mathbf{a}_p$ is the vector of primary mode amplitudes and $\mathbf{B}$ is the matrix of coupling coefficients.

98 CHAPTER 4. ADAPTIVE CONTROL OF SOUND RADIATION

As detailed in Section 2.6.1, the sound power radiated from a vibrating panel is related to the amplitudes of the structural modes by the expression

$$\Pi = \sum_{i=1}^{N} \sum_{j=1}^{N} \hat{M}_{ij} a_i^* a_j \qquad (4.3)$$

and so the radiated sound power can be written in compact form as

$$\Pi = \mathbf{a}^H \hat{\mathbf{M}} \mathbf{a} \qquad (4.4)$$

where the superscript ^{H} denotes the Hermitian (conjugate transpose) and $\hat{\mathbf{M}}$ is the real, symmetric and positive definite radiation efficiency matrix, defined in Eq. (2.67). Its diagonal terms are the self-radiation resistances, which are proportional to the self-radiation efficiencies of the structural modes, and the off-diagonal terms are the mutual-radiation resistances, which are proportional to the combined radiation efficiencies due to the interaction between different structural modes.

By substituting Eq. (4.2) into Eq. (4.4), a quadratic function is obtained. It has a unique minimum when the complex amplitudes of the secondary forces are equal to

$$\mathbf{f}_{s,opt,\Pi} = -\left[\mathbf{B}^H \hat{\mathbf{M}} \mathbf{B}\right]^{-1} \hat{\mathbf{M}} \mathbf{B}^H \mathbf{a}_p. \qquad (4.5)$$

As a result, structural sensors, such as shaped PVDF strips, are suitable for capturing sound radiation from vibrating structures. An optimal set of secondary forces can thus be generated and superimposed to the unwanted (primary) disturbance so as to achieve noise reduction at and in the vicinity of the error point. However, a considerable number of spatially distributed sensors, bonded to or integrated into the structure, may be necessary to capture the radiation characteristics of the panel, especially at high frequencies. Moreover, the direct computation of the optimal secondary sources may be computationally demanding because of the matrix inversion.

Alternatively, the radiated sound power can be estimated by using an array of microphones positioned at a certain distance from the radiating panel in order to avoid near-field effects. A smaller number of microphones may be sufficient, especially if pressure and pressure gradient are measured simultaneously. This technique, known as sound intensimetry, will be detailed in Chapter 6.

For cavity control of multi-panel partitions, the total acoustic potential energy, measured in the gap between the vibrating panels, represents an efficient control objective. This strategy is particularly effective when the

4.2. CONTROL OF SOUND RADIATION

radiating panel exhibits high modal density. In this case, the cost function can be written as

$$\Xi(\omega) = \frac{1}{4\rho_0 c^2} \int_V |p(x,y,z,\omega)|^2 \, dV \qquad (4.6)$$

where $p(x,y,z,\omega)$ is the complex acoustic pressure at the point (x,y,z) in the acoustic cavity.

This function evaluates the sum of the square outputs in a number of pressure microphones placed within the enclosure. For a harmonically excited enclosed sound field, the complex acoustic pressure can be represented by a finite number $(\hat{N})$ of modal contributions. From the modal orthogonality property, it follows that

$$\Xi(\omega) = \frac{V}{4\rho_0 c^2} \sum_{n=0}^{\hat{N}} |\hat{a}_n(\omega)|^2 = \frac{V}{4\rho_0 c^2} \hat{\mathbf{a}}^H \hat{\mathbf{a}} \qquad (4.7)$$

where the coefficients $\hat{a}_n$ are the acoustic mode amplitudes.

Similarly to Eq. (4.2), the mode amplitude vector $\hat{\mathbf{a}}$ can be written in the form

$$\hat{\mathbf{a}} = \hat{\mathbf{a}}_p + \hat{\mathbf{B}} \mathbf{q}_s \qquad (4.8)$$

where M secondary sources of complex strengths q_{sm} are introduced in order to control the primary acoustic field.

By substituting Eq. (4.8) into Eq. (4.7), it follows that the total acoustic potential energy Ξ can be expressed in the Hermitian quadratic form as

$$\Xi(\omega) = \frac{V}{4\rho_0 c^2} \left[\mathbf{q}_s^H \hat{\mathbf{B}}^H \hat{\mathbf{B}} \mathbf{q}_s + \mathbf{q}_s^H \hat{\mathbf{B}}^H \hat{\mathbf{a}}_p + \hat{\mathbf{a}}_p^H \hat{\mathbf{B}} \mathbf{q}_s + \hat{\mathbf{a}}_p^H \hat{\mathbf{a}}_p \right]. \qquad (4.9)$$

If the number of acoustic modes significantly excited by the primary source is higher than the secondary sources, matrix $\hat{\mathbf{B}}$ is not square. The optimal set of secondary source strength which minimizes the quadratic function of Eq. (4.9) is given by

$$\mathbf{q}_{s,opt,\Xi} = -\left[\hat{\mathbf{B}}^H \hat{\mathbf{B}}\right]^{-1} \hat{\mathbf{B}}^H \hat{\mathbf{a}}_p. \qquad (4.10)$$

In order to achieve the best possible reduction, an infinite number of sensors would be necessary to estimate Ξ. Nevertheless, apart from minimizing a global cost function, such as the radiated sound power or the total acoustic potential energy, an active control system can also be designed to minimize the structural or acoustic responses at discrete points. In practice, one is restricted to using a finite number of measurements points, usually referred to as error sensors, in order to estimate a global cost function. A method for producing uniform reductions in the radiated sound field is given by the minimization of the sum of the squares of the pressure signals measured at L error locations. The resulting cost function is given by the sum of the squared error signals

$$J = \sum_{l=1}^{L} |e_l|^2 = \mathbf{e}^H \mathbf{e}. \qquad (4.11)$$

If the disturbance is assumed to be tonal and of frequency $\omega_0 T$ and the system under control is linear, active control can be performed by driving a number of control actuators to suppress the disturbance over a given frequency range. This requires the observed primary and secondary signals at each error point to be equal in magnitude and to have a phase difference of 180°. The complex response at each microphone can be expressed in matrix form as the response due to a linear superposition of the primary disturbance signals and M individual secondary control signals,

$$\mathbf{e}(e^{j\omega_0 T}) = \mathbf{d}(e^{j\omega_0 T}) + \mathbf{G}(e^{j\omega_0 T})\mathbf{u}(e^{j\omega_0 T}) \qquad (4.12)$$

where $\mathbf{G}$ is the matrix of complex transfer responses between each secondary control signal and the error microphones, $\mathbf{u}$ is the vector of complex inputs to the secondary control actuators and $\mathbf{d}$ is the vector of complex primary excitation. Substituting Eq. (4.12) into Eq. (4.11) yields the Hermitian quadratic form

$$J = \mathbf{u}^H \mathbf{G}^H \mathbf{G} \mathbf{u} + \mathbf{u}^H \mathbf{G}^H \mathbf{d} + \mathbf{d}^H \mathbf{G} \mathbf{u} + \mathbf{d}^H \mathbf{d}. \qquad (4.13)$$

If the overdetermined case is considered, in which more sensors than actuators are involved, the resulting optimal control vector is

$$\mathbf{u}_{opt} = -\left[\mathbf{G}^H \mathbf{G}\right]^{-1} \mathbf{G}^H \mathbf{d} \qquad (4.14)$$

and substituting this expression into Eq. (4.13) gives the expression for the minimum value

$$J_{min} = \mathbf{d}^H \left[\mathbf{I} - \mathbf{G}\left[\mathbf{G}^H \mathbf{G}\right]^{-1} \mathbf{G}^H\right] \mathbf{d}. \qquad (4.15)$$

4.3 Control Strategies

In this section, the application of feedforward and feedback control to active suppression of noise radiation is briefly reviewed. In either case, the controller, which is the *brain* of the control system, takes in sensor signals, calculates the appropriate control action and drives control sources in order to realize the control objective. The technical literature on control theory and methodologies is enormous, and it is well beyond the scope of this technical book. The two strategies are explained in their simplest terms for SISO linear systems and attempts are made to generalize these approaches rather than to give an exhaustive list of them. The strategies are extensible to MIMO systems.

4.3.1 *Feedforward control*

Feedforward control generally refers to those applications where the detection sensors, which contain the relevant information about the sound field to be cancelled, are fed through the controller to produce the outputs to the control sources. The input to the controller is distinct from the residual error, which is to be attenuated. The block diagram of a general feedforward controller is presented in Fig. 4.3. The basic assumption behind a pure open-loop feedforward control is the time invariance of the system to be controlled. The vector $\mathbf{d}(k)$ denotes the disturbance to be suppressed while it is assumed that a reference signal $\mathbf{x}(k)$ which is correlated with the disturbance can be measured. The reference signal is thus assumed to describe the source which causes the disturbance $\mathbf{d}(k)$. The feedforward control thus consists of transforming the reference signal directly into an

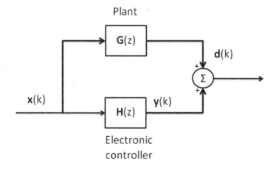

Figure 4.3. Block diagram of a general feedforward control

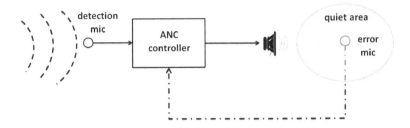

Figure 4.4. Feedforward control strategy for active noise control

input to the control actuator. Interference of $\mathbf{y}(k)$ with the disturbance $\mathbf{d}(k)$ is aimed at suppression of the latter. In linear control theory, the feedforward controller $\mathbf{H}(z)$ is then determined off-line such that it provides the appropriate magnitude and phase relation between the reference signal $\mathbf{x}(k)$ and the control signal $\mathbf{y}(k)$, which yields maximum suppression.

The performance of a feedforward control system is very sensitive to the slightest changes in $\mathbf{G}(z)$, due to changes in the dynamics of the acoustic system to be controlled. They may be caused, for instance, by temperature or pressure variations in aircraft interior or as a consequence of passengers moving around. In order to cope with time varying systems, additional sensors located in the area to be silenced may be used to adapt the controller to the changing environment. As shown in Fig. 4.4, the performance of a feedforward controller can be augmented with a measure of the system response. The following paragraphs give an overview of different methods to the iterative optimization of the adaptive feedforward control parameters.

4.3.2 Feedback control

When no reference signal is available that is well correlated with the disturbance, then the radiation error sensor may be used to drive the secondary sources via a feedback controller. A feedback control system is shown in Fig. 4.5 while a block diagram representation is given in Fig. 4.6. The plant $G(s)$ is the complex transfer function between the control actuator input and the residual sensor output as a function of $s = j\omega$.

The reduction in response due to the feedback control loop is given by the transfer function:

$$S(s) = \frac{E(s)}{D(s)} = \frac{1}{1 + G(s)H(s)}. \quad (4.16)$$

4.3. CONTROL STRATEGIES

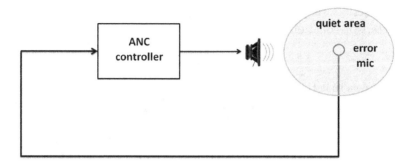

Figure 4.5. Feedback control strategy for active noise control

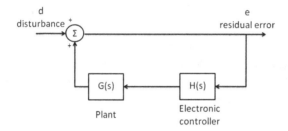

Figure 4.6. Block diagram of a general feedback control

This transfer function, also known as sensitivity function, needs to be small if the objective of the control system is the rejection of the disturbance.

In order to make the control response small compared to the uncontrolled response it is necessary to make

$$|GH| \gg 1. \tag{4.17}$$

The transfer function GH is termed the loop gain. Thus the first principle of feedback control is to make the loop gain GH large or equivalently

$$|H| \gg |G^{-1}| \tag{4.18}$$

over the frequency band where noise attenuation is desired. Notice that in the regulation band where $|GH| \gg 1$, the phase of $G(s)H(s)$ is immaterial to the performance.

The stability requirements impose particular conditions between the amplitude and phase of the loop gain. The poles of the transfer function which describes the system shown in Fig. 4.6 are given by the roots of the characteristic equation

$$1 + G(s)H(s) = 0. \tag{4.19}$$

It follows that the stability of a feedback system is met only if the poles of the closed-loop transfer function are in the open left-hand half of the s plane.

There are a wide variety of methods for designing a feedback control system. Generally speaking, increased control bandwidth leads to increased system complexity and cost for typical acoustic spaces or structures. However, in practical implementations, the plant may introduce significant phase shift due to the secondary source mechanical resonances. The electroacoustic resonance frequencies of a loudspeaker, for instance, may fall below 100 Hz. Moreover, the acoustic path from the secondary acoustic source to the microphone will also inevitably involve some delays due to the acoustic propagation time. Consequently, as the phase shift in the plant approaches 180°, the negative feedback action becomes positive feedback and the control system can become unstable.

4.3.3 Feedforward vs. feedback control

In order to better assess the two control strategies, the layout of a feedforward and feedback control system for active control of sound transmission through double-glazing windows is following discussed. In both cases, the minimization on the sound transmission loss results from the use of loudspeakers inside the cavity between the two panels. Such a control layout, usually referred to as a cavity control in multi-wall partitions control theory, is based on the active control of the cavity resonances. A literature review of cavity control is given in Section 5.2.2. The difference between the two control configurations is the generation of the reference signal. In a pure feedforward architecture, the reference may be obtained either directly from the signal generator or from a microphone in the sending room of the testing facility. Feedback control may be realized through a number of methods ranging from PID-control design for SISO systems to model-based control techniques. Figures 4.7 and 4.8 show a feedforward set-up and a feedback set-up, respectively. The latter highlights how a feedback control problem can always be transformed into an apparent feedforward one with the advantage of being easily implemented on a DSP as its feedforward counterpart. The controller may use the error signal and the counteracting

4.4. STEEPEST-DESCENT ALGORITHM

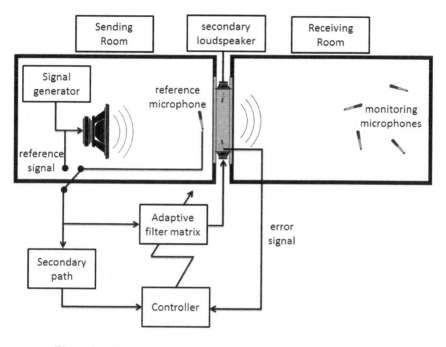

Figure 4.7. Test set-up with feedforward controller for cavity control

signal in order to estimate the disturbance, which serves as reference signal for the adaptive algorithm.

Due to the difficulties in designing a stable feedback controller, an alternative practice is to place the secondary sources close to the error sensors in a collocated feedback control arrangement. As always in the case of adaptive active noise control, the achievable performance depends upon the nature of the signals. The measured transmission loss may be affected by several factors, such as a feedback from the loudspeaker inside the cavity to the reference microphone which may not be compensated, the reference microphone location which may correspond to some nodes of the sending room modes, etc. However, the latter may be *a priori* evaluated by measuring the transfer function between primary and reference signals.

4.4 Steepest-Descent Algorithm

Optimum cancellation of noise disturbance can be realized by detecting noise and processing this by a suitable electronic controller. However, in practice, since the spectral contents of these disturbances as well as the

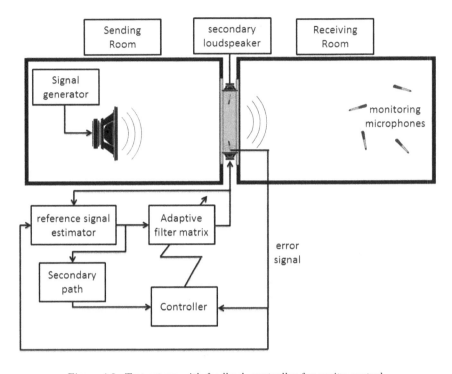

Figure 4.8. Test set-up with feedback controller for cavity control

characteristics of system components are in general subject to variation, the controller is further required to be adaptive in order to track these changes and maintain the desired performance. In realizing such a need, the steepest descent algorithm can be used to iteratively approach the optimal controller by adjusting the signals driving the secondary control actuators in order to minimise the sum of the squared moduli of the outputs given by a number of error sensors. This approach has the additional advantage that it can also compensate, at the same time, any possible nonlinearities in the system under control.

Introducing the derivatives of the cost function J with respect to the real and imaginary part of the vector of complex inputs to the secondary sources, $\mathbf{u} = \mathbf{u}_R + j\mathbf{u}_I$, the method of steepest descent, at the nth iteration, reads

$$\mathbf{u}_R(n) = \mathbf{u}_R(n-1) - \mu \frac{\partial J}{\partial \mathbf{u}_R(n-1)} \qquad \mathbf{u}_I(n) = \mathbf{u}_I(n-1) - \mu \frac{\partial J}{\partial \mathbf{u}_I(n-1)} \tag{4.20}$$

where μ is a convergence coefficient.

4.5. ADAPTIVE DIGITAL FILTERS

The two expressions can be combined into a single complex form such that

$$\mathbf{u}(n+1) = \mathbf{u}(n) - \mu \mathbf{g}(n) \tag{4.21}$$

where $\mathbf{g}$ is the complex gradient vector, which is defined as

$$\mathbf{g} = \frac{\partial J}{\partial \mathbf{u}_R} + j \frac{\partial J}{\partial \mathbf{u}_I}. \tag{4.22}$$

For the Hermitian quadratic function, given by Eq. (4.13), the complex gradient vector $\mathbf{g}$ at the nth iteration is equal to

$$\mathbf{g}(n) = 2 \left(\mathbf{G}^H \mathbf{G} \mathbf{u}(n) + \mathbf{G}^H \mathbf{d} \right). \tag{4.23}$$

Assuming that the vector of complex error signals at the nth iteration can be written as $\mathbf{e}(n) = \mathbf{d} + \mathbf{G} \mathbf{u}(n)$, the steepest-descent algorithm which minimizes the sum of the squared error signals is given by

$$\mathbf{u}(n+1) = \mathbf{u}(n) - \alpha \mathbf{G}^H \mathbf{e}(n) \tag{4.24}$$

where α is a convergence coefficient equal to 2μ.

The steepest-descent algorithm is widely used in practical applications for the robust control of sound fields. Its implementation on digital control systems can be particularly effective in reducing sound radiation from structures, especially at low frequencies.

4.5 Adaptive Digital Filters

Controlling a physical system with a sampled data controller requires the manipulation of continuous time signals by digital filters. In a linear, causal, digital system, the output signal $y(n)$ at the time step n depends on the current and past values of the input signal $x(n)$. It can be written as

$$y(n) = \sum_{i=0}^{\infty} h_i x(n-i) \tag{4.25}$$

with h_i being the samples of the impulse response of the system. In order to implement a practical digital filter, the infinite summation is truncated after I samples to give at each output sample

$$y(n) = \sum_{i=0}^{I-1} w_i x(n-i) \tag{4.26}$$

where w_i are the coefficients or weights of the digital filter of Ith order.

108 CHAPTER 4. ADAPTIVE CONTROL OF SOUND RADIATION

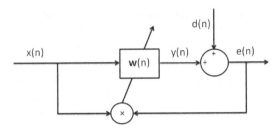

Figure 4.9. Representation of a FIR filter whose coefficients are adapted using the LMS algorithm

Consider the single-channel feedforward control problem shown in the block diagram of Fig. 4.9. The measured error signal $e(n)$ can be expressed as the sum of the disturbance signal $d(n)$, and the control signal $y(n)$.

$$e(n) = d(n) + y(n) = d(n) + \sum_{i=0}^{I-1} w_i x(n-i). \quad (4.27)$$

If the control is extended to stochastic disturbances, the finite impulse response (FIR) filter coefficients w_i shall minimize the expected value of the sum of the squared error signals, independently at each frequency. This error criterion can thus be expressed as:

$$J = E[e^2(n)] \quad (4.28)$$

where E [·] denotes the expectation operator. If all the signals are stationary and ergodic, the expectation is time invariant and can be calculated by averaging over time. The cost function is therefore equal to the average mean-square value of the error signal. The value of the coefficients of the FIR filter that reduces the mean-square error to a minimum can be found by differentiating J with respect to each filter coefficient and setting the derivatives to zero. The filter that has these optimal coefficients $\mathbf{w}_{opt}$ is known as the Wiener filter [93]. The optimal filter coefficients vector depends on the auto- and cross-correlation functions of the reference and disturbance signals.

As introduced in Section 4.4, an alternative approach to determining the coefficients of such a filter is to make them adaptive. The coefficients may be iteratively adjusted so that the cost function J is minimized by moving towards the global minimum of the error surface. In a steepest

4.5. ADAPTIVE DIGITAL FILTERS

descent algorithm, the adaptation is given by

$$\mathbf{w}(n+1) = \mathbf{w}(n) - \mu \frac{\partial J}{\partial \mathbf{w}}(n) \tag{4.29}$$

where

$$\frac{\partial J}{\partial \mathbf{w}} = 2E\left[\mathbf{x}(n)e(n)\right]. \tag{4.30}$$

However, instead of updating the filter coefficients with an averaged estimate of the gradient, the coefficients can be updated at every sample by using an instantaneous estimate of the gradient. This quantity, which is known as stochastic gradient, is equal to the derivative of the instantaneous error with respect to the filter coefficients

$$\frac{\partial e^2}{\partial \mathbf{w}} = 2 \cdot \mathbf{x}(n)e(n). \tag{4.31}$$

The adaptation algorithm for the filter coefficients thus becomes

$$\mathbf{w}(n+1) = \mathbf{w}(n) - \alpha \cdot \mathbf{x}(n)e(n). \tag{4.32}$$

This algorithm, known as the least mean squares (LMS) algorithm, is widely used in several noise control applications because of its numerical simplicity and robustness. The error signal is multiplied by the reference signal to provide a cross-correlation estimate used to adapt the filter coefficients. By applying the LMS algorithm to a feedforward problem, the maximum achievable attenuation at the error signal depends upon two factors. First, the degree of coherence between the reference and the disturbance signals, and second, the ability of the controller to implement the frequency response required for the optimal control. The algorithm converges for $0 < \alpha < 2/(I \cdot \bar{x}^2)$ where $\bar{x}^2$ is the mean square value of $x(n)$ and $\alpha = 2 \cdot \mu$ [93].

In a practical feedforward problem, however, the adaptive controller must take the secondary path transfer function into account. As shown in Fig. 4.10, the dynamic system, including the actuators and the amplifiers between the output of the controller and the observation point, can lead to a distortion of the cross-correlation estimate. Consequently, for the adaptive control algorithm to be stable, an internal model of the plant response can be required. Furthermore, an adaptive task has to be performed when the response of the system to a given secondary excitation changes with time. In this case, the adaptation has to be modified in order to simultaneously ensure both control and identification.

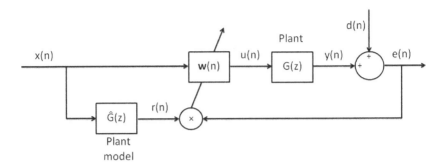

Figure 4.10. Block diagram of the filtered-reference LMS algorithm

4.6 Filtered-X LMS Algorithm

Figure 4.10 shows the block diagram of an active control system implementing a filtered-X LMS algorithm to track time-varying disturbances by controlling the dynamic behaviour of the plant $G(z)$. The reference signal is prefiltered by an *estimated* version of the true plant response G in order to give a valid cross-correlation estimate of the measured error signal and the reference signal. The internal plant can be implemented as a separate real-time filter, $\hat{G}$, from which the filtered reference signal, $\mathbf{r}(n)$, is derived. It should be noticed that when implemented on a digital processor, the exact secondary path G is not known but only a measured estimate of it, denoted as $\hat{G}$.

The plant G, between the output of the control filter and the error point, is rarely constant over time, minimum phase and hardly known with perfect precision. This complicates the design of an adaptive algorithm giving rise, in the simplest case, to a pure delay in the response that could destabilize the control algorithm. By assuming that the plant G can be described by a FIR filter of order M, the error signal can be written as

$$e(n) = d(n) + y(n) = d(n) + \sum_{j=0}^{M-1}\sum_{i=0}^{I-1} g_j w_i(n-j) x(n-j-i) \quad (4.33)$$

where g_j is the impulse response of the physical plant whose z transform is $G(z)$.

The principle of the *filtered-reference LMS algorithm* (FX-LMS) is to adapt the filter in the opposite direction to the instantaneous gradient of the mean square error with respect to the filter coefficients. In this case,

4.7. MULTI-CHANNEL ADAPTIVE ALGORITHMS

the instantaneous gradient can be written as

$$\frac{\partial e^2(n)}{\partial \mathbf{w}} = 2 \cdot e(n) \cdot \frac{\partial e(n)}{\partial \mathbf{w}} = 2 \cdot e(n) \cdot \mathbf{r}(n) \quad (4.34)$$

where $\mathbf{r}(n) = [r(n), \ldots, r(n-I+1)]^T$ is the vector of past values of the filtered reference signal.

In comparison to Eq. (4.31), the reference signal is filtered by the plant model $\hat{G}$. Therefore, the adaptation algorithm reads as follows:

$$\mathbf{w}(n+1) = \mathbf{w}(n) - \alpha \mathbf{r}(n) e(n) \quad (4.35)$$

where $\alpha = 2\mu$ is the convergence coefficient. Although the filtered-X LMS algorithm is motivated by the control of stochastic disturbances, it can provide a very efficient method for adapting the controller in case of a narrow-band time-varying disturbance.

4.7 Multi-Channel Adaptive Algorithms

With the advent of digital signal processors (DSPs), control complexity has rapidly grown from low-order SISO systems, often realized in analog circuits, to high-order, adaptive, MIMO digital controllers driving a plurality of control sources based on information gained by a multitude of sensor signals.

The multi-channel version of the filtered-X LMS algorithm is called the multiple error LMS algorithm (MELMS) [93]. It is assumed that a number of K reference signals, M secondary sources and L error sensors are available. The block diagram of the MIMO feedforward control architecture being considered in this section is shown in Fig. 4.11. Block C represents a matrix of $L \times M$ error paths representing the transfer functions from each secondary source to each error sensor. Block W is a matrix of $K \times M$

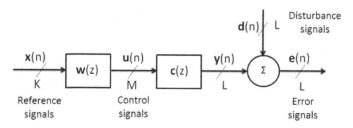

Figure 4.11. Multi-channel feedforward control with M control sources, K reference signals and L error sensors

control filters that are designed to minimize the sum of the squares of L error signals. The system under control is considered time-invariant. The output of the lth error sensor can be written as

$$e_l(n) = d_l(n) + \sum_{k=1}^{K}\sum_{m=1}^{M}\sum_{j=0}^{J-1} c_{lmj} \sum_{i=0}^{I-1} w_{mki} x_k(n-i-j) \qquad (4.36)$$

where $d_l(n)$ is the primary noise at the lth error sensor, c_{lmj} is the jth coefficient of the impulse response from the mth secondary source to the lth error sensor. The signal driving the mth actuator is calculated from K reference signals, each filtered by the Ith order FIR control filter with coefficient w_{mki}.

Equation (4.36) can also be expressed in matrix form as

$$\mathbf{e}(n) = \mathbf{d}(n) + \mathbf{R}(n)\mathbf{w} \qquad (4.37)$$

where $\mathbf{R}(n)$ is the filtered reference signal matrix and

$$\mathbf{e}(n) = [e_1(n), e_2(n), \ldots, e_L(n)]^T \qquad (4.38)$$
$$\mathbf{d}(n) = [d_1(n), d_2(n), \ldots, d_L(n)]^T \qquad (4.39)$$
$$\mathbf{w} = [\mathbf{w}_0^T, \mathbf{w}_1^T, \ldots, \mathbf{w}_{I-1}^T]^T \qquad (4.40)$$

with

$$\mathbf{w}_i = [w_{11i}, w_{12i}, \ldots, w_{1Ki}, w_{21i}, \ldots, w_{MKi}]^T \qquad (4.41)$$

and

$$\mathbf{R}(n) = \begin{bmatrix} \mathbf{r}_1^T(n) & \mathbf{r}_1^T(n-1) & \ldots & \mathbf{r}_1^T(n-I+1) \\ \mathbf{r}_2^T(n) & \mathbf{r}_2^T(n-1) & \ldots & \mathbf{r}_2^T(n-I+1) \\ \vdots & & & \\ \mathbf{r}_L^T(n) & \mathbf{r}_L^T(n-1) & \ldots & \mathbf{r}_L^T(n-I+1) \end{bmatrix} \qquad (4.42)$$

with

$$\mathbf{r}_l(n) = [r_{l11}(n), r_{l12}(n), \ldots, r_{l1K}(n), r_{l21}(n), \ldots, r_{lMK}(n)]^T. \qquad (4.43)$$

The cost function J is given by the expectation of the sum of the square error signals

$$J = E\left[\sum_{l=1}^{L} e_l^2(n)\right] = E\left[\mathbf{e}^T \mathbf{e}\right] = \mathbf{w}^T E\left[\mathbf{R}^T \mathbf{R}\right] \mathbf{w} + 2\mathbf{w}^T E\left[\mathbf{R}^T \mathbf{d}\right] + E\left[\mathbf{d}^T \mathbf{d}\right]. \qquad (4.44)$$

In Eq. (4.44), the quadratic nature of the error surface is clearly identified. This function of the controller filter coefficients $\mathbf{w}$ has a unique global

4.7. MULTI-CHANNEL ADAPTIVE ALGORITHMS

minimum if $E\left[\mathbf{R}^T\mathbf{R}\right]$ is positive definite. By setting the differential of this expression with respect to $\mathbf{w}$ to zero, the optimum Wiener set of coefficients may be obtained. It is equal to

$$\mathbf{w}_{opt} = -\left\{E\left[\mathbf{R}^T\mathbf{R}\right]\right\}^{-1} E\left[\mathbf{R}^T\mathbf{d}\right]. \tag{4.45}$$

This set of filter coefficients gives a minimum error criterion equal to

$$J_{\min} = J_0 - E\left[\mathbf{d}^T\mathbf{R}\right]\left\{E\left[\mathbf{R}^T\mathbf{R}\right]\right\}^{-1} E\left[\mathbf{R}^T\mathbf{d}\right] \tag{4.46}$$

where $J_0 = E[\mathbf{d}^T\mathbf{d}]$ is the value of the error criterion with no control applied.

In common with the single channel filtered reference algorithm, the control filter coefficients of the multi-channel algorithm may be adjusted iteratively at every sample time by an amount proportional to the negative instantaneous value of the stochastic estimation of the gradient vector. The vector of derivatives of the instantaneous squared errors with respect to the control coefficients is given by

$$\frac{\partial \mathbf{e}^T(n)\mathbf{e}(n)}{\partial \mathbf{w}(n)} = 2\mathbf{R}^T(n)\mathbf{e}(n). \tag{4.47}$$

Equation (4.47) is therefore used to adapt the control filter coefficients as

$$\mathbf{w}(n+1) = \mathbf{w}(n) - \alpha \mathbf{R}^T(n)\mathbf{e}(n) \tag{4.48}$$

where α is a convergence coefficient. An interesting convergence analysis of the MELMS algorithm is reported in [117].

For large multi-channel systems, the filtered-X operations of the reference signals require intensive computations. An alternative method for reducing the computational load of the multi-channel filtered-X LMS algorithm consists of using only one error signal at each algorithm iteration to adapt all the filter coefficients. In the least maximum mean squares (LMMS) algorithm, the stochastic cost function consists of the instantaneous maximum value of the error signals as follows

$$J(n) = \max_{1 \leq l \leq L}\left\{(e_l(n))^2\right\} = (e_b(n))^2. \tag{4.49}$$

The update weights equation is therefore given by

$$\mathbf{w}(n+1) = \mathbf{w}(n) - \alpha \mathbf{R}_b^T(n)e_b(n) \tag{4.50}$$

where subscript b denotes the error signal with maximum squared value for a given value of the control vector. At any one instant, the algorithm

Table 4.1 Computational operations per sample in a $1 \times M \times L$ multi-channel system using the MELMS, LMMS and scanning error LMS algorithms [116]

	MELMS	LMMS	Scanning
Multiplications	$((JI+1)L+1)M$	$(JI+2)M+3L$	$(JI+2)M$
Additions	$((J-1)IL+1)M$	$((J-1)I+1)M+2L$	$((J-1)I+1)M$

is similar to that obtained by minimizing the sum of the squared levels at all error sensors, but it can be evaluated with a lower computational effort. On the other hand, a comparison between the error signal magnitudes has to be carried out and the error signal with the highest level may generally switch from one sensor to another during the course of adaptation. Thus the minimization of the maximum level at any sensor will result in a controlled field that is more spatially uniform.

Another method for reducing the computation load is the scanning error LMS algorithm. In this case, the error signal is chosen in turn at each algorithm iteration to give

$$w_{mki}(n+1) = w_{mki}(n) - \alpha e_1(n) r_{1mk}(n-i)$$
$$w_{mki}(n+2) = w_{mki}(n+1) - \alpha e_2(n+1) r_{2mk}(n-i+1)$$
$$\ldots$$
$$w_{mki}(n+L) = w_{mki}(n+L-1) - \alpha e_L(n+L-1) r_{Lmk}(n-i+L-1). \quad (4.51)$$

The computational complexity of multi-channel systems using the algorithms mentioned above is compared in Table 4.1. Unlike the MELMS and LMMS, the scanning error algorithm computational load does not depend on the number of error sensors, thus requiring much fewer operations per sampling period. For few error sensors, the computational load of the LMMS is comparable to that of the scanning error LMS.

4.8 Experimental System Identification

The adaptive control algorithms described above require a very precise model of the system to perform well. In general, an accurate knowledge of the system under control can be derived from numerical simulations and/or deduced from experimental measurements. The model can be numerical,

4.8. EXPERIMENTAL SYSTEM IDENTIFICATION

and should account for all the features of the real system, such as nonlinearities, temperature effects, etc. Alternatively, the model can be experimental, previously-recorded, and stored in the control algorithm.

It is often the case, when implementing an actual active control system, that an experimentally identified model is required for a correct operation. It has the advantage of being more precise for complicated systems and it can be reevaluated during the lifetime if any change occurs in the environment or the system itself. With the recent advances in computer technology for data acquisition, signal processing and analysis, plant models can be identified accurately from the measured responses and excitations using system identification techniques. Moreover, if the response of the plant does not change significantly over the timescale of operation of the control system, the system identification can be performed off-line.

In the reference application, the noise transmission path from the exterior to the interior of an aircraft sidewall panel is simplified through an experimental arrangement aimed at characterizing the sound transmission properties of the smart window prototype. This experimental model refers to the set-up that will be detailed in Chapter 6. The system schematic is presented in Fig. 4.12. It consists of two main parts: a soundproof box and the triple wall test-bed mounted on the side opening of the box. The incident acoustic field in this application is the noise generated by the propeller of a turboprop aircraft approximated by acoustical oblique incident sound waves. Such a primary noise source is generated by a loudspeaker placed inside the box. The radiated acoustic field of the window panels corresponds to the aircraft interior which is approximated by an acoustical free field. The experimental characterization is performed within a semi-anechoic chamber. In this way, the sound power transmitted through the structure can be determined by integrating the radiated sound pressure in the far field over a surface surrounding the radiating plate. Control is provided by piezoelectric actuators mounted on the radiating plate of the triple panel system. The structural inputs are determined by the filtered-reference LMS control theory. A number of microphones are used to provide either the error signals for the adaptive controller or the sound pressure level attenuations.

Considering the experimental test-bed used for the real-time control, an experimental model-based approach is used. The experimental models of the unknown transfer functions, relating the input and output of the system under control, are recorded off-line, before the control system operation. The plant responses are measured for both the disturbing noise source and

116 CHAPTER 4. ADAPTIVE CONTROL OF SOUND RADIATION

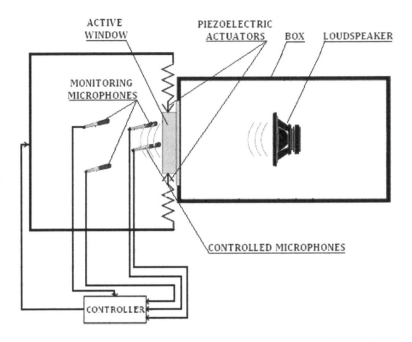

Figure 4.12. Sketch of the active structural acoustic control of the triple panel system

the active window prototype driven by the control actuators. Next, they are numerically fitted and integrated in the active noise control algorithm. The acoustic field emitted by the loudspeaker is characterized in the 50–800 Hz frequency range. A broadband signal is used. Four microphones are placed in the receiving room in order to measure the corresponding transfer functions. The sensor positions are listed in Table 6.2.

For a linear MIMO system, the transfer function between the input and the output can be expressed as:

$$\begin{pmatrix} Y_1 \\ Y_2 \\ Y_3 \\ .. \\ Y_n \end{pmatrix} = \begin{pmatrix} \frac{N_{Y_{11}}}{D_{Y_{11}}} & \frac{N_{Y_{12}}}{D_{Y_{12}}} & \frac{N_{Y_{13}}}{D_{Y_{13}}} & .. & \frac{N_{Y_{1m}}}{D_{Y_{1m}}} \\ \frac{N_{Y_{21}}}{D_{Y_{21}}} & \frac{N_{Y_{22}}}{D_{Y_{22}}} & \frac{N_{Y_{23}}}{D_{Y_{23}}} & .. & \frac{N_{Y_{2m}}}{D_{Y_{2m}}} \\ \frac{N_{Y_{31}}}{D_{Y_{31}}} & \frac{N_{Y_{32}}}{D_{Y_{32}}} & \frac{N_{Y_{33}}}{D_{Y_{33}}} & .. & \frac{N_{Y_{3m}}}{D_{Y_{3m}}} \\ .. & .. & .. & .. & .. \\ \frac{N_{Y_{n1}}}{D_{Y_{n1}}} & \frac{N_{Y_{n2}}}{D_{Y_{n2}}} & \frac{N_{Y_{n3}}}{D_{Y_{n3}}} & .. & \frac{N_{Y_{nm}}}{D_{Y_{nm}}} \end{pmatrix} \begin{pmatrix} U_1 \\ U_2 \\ U_3 \\ .. \\ U_m \end{pmatrix} \quad (4.52)$$

where the ith transfer function can be expanded as follows:

$$\frac{N_{Y_i}(s)}{D_{Y_i}(s)} = \frac{b_{0i} + b_{1i}s + b_{2i}s^2 + b_{3i}s^3 + \cdots + b_{ki}s^k}{a_{0i} + a_{1i}s + a_{2i}s^2 + a_{3i}s^3 + \cdots + a_{pi}s^p}. \quad (4.53)$$

4.8. EXPERIMENTAL SYSTEM IDENTIFICATION

The numerator and denominator expressions are polynomial functions in the Laplace variable, s. A strictly proper transfer function is obtained if the highest order of the numerator polynomial (k) is at least one order lower than the highest order of the denominator polynomial (p).

The fitting of the experimental curves is realized by using the `invfreqs` function from the MATLAB® Signal Processing Toolbox. This function allows to compute the best coefficients b_i and a_i of the polynomial function defined in Eq. (4.53), specifying the number of states of the numerator, k, and of the denominator, p. Each transfer function is processed by using an error function defined as:

$$f(n_p, n_k) = ||H_{exp} - H_{fit}||^2 \tag{4.54}$$

where n_p and n_k are respectively the number of states of the numerator and the denominator of the transfer function and $||H_{fit} - H_{exp}||$ is the Euclidean norm of the difference between the experimental frequency response function (FRF) and the fitted transfer function. The coefficients are computed from experimental data and then converted in a canonical form of state-space model as follows:

$$\begin{aligned}\dot{\mathbf{x}} &= \mathbf{Ax} + \mathbf{Bu} \\ \mathbf{y} &= \mathbf{Cx} + \mathbf{Du}.\end{aligned} \tag{4.55}$$

The state-space models computed for the primary and secondary sources are then used for the active noise control system design. A perfect correspondence between the modelled plant $\hat{G}(z)$ and the real plant $G(z)$ is assumed for the numerical simulations. Moreover, despite the uncertainties in the modelling errors, the off-line models of the error signal paths $\hat{G}(z)$ will be included in the real-time controller described in Chapter 6.

Since the multi-channel off-line experimental modelling is a difficult task due to the acoustical coupling between the individual channels, only one error path from either the loudspeaker or the control actuators to one of the microphones is estimated at a time. The acoustic transfer functions between the input stimulus (voltage) and the resulting sound field are experimentally characterized inside the anechoic chamber. Figure 4.13 shows a detail of the acoustic responses measured from the input of the loudspeaker acting as noise disturbance to the output from the pressure microphones inside the receiving room. Band limited white noise is used as noise disturbance. The identification process is evaluated by comparing the experimental and the fitted transfer functions measured at each error microphone. The simulation results in Fig. 4.13 show a perfect agreement in magnitude and

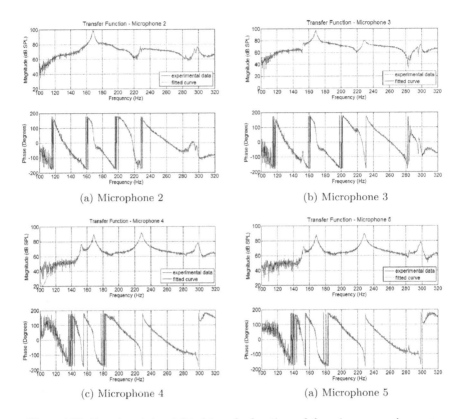

Figure 4.13. Experimental and fitted transfer functions of the primary sound source

phase between the numerical and experimental data. Small deviations in amplitudes are obtained at low-frequency at microphones 4 and 5. Although the acoustic field in the semi-anechoic chamber is sufficiently uniform due to its flat acoustic response, the measured frequency response functions are influenced by the vibroacoustic response of the sending room. This aspect will be addressed in Chapter 6.

4.9 Numerical Simulations

This section theoretically investigates the active noise reduction that can be achieved with a predetermined configuration of secondary sources mounted on the smart panels using a feedforward control scheme. The aim is, firstly, to show the potential of active control to improve the sound insulation

4.9. NUMERICAL SIMULATIONS

of triple panel systems and, secondly, to give a comprehensive overview of the physical system under investigation. Sound radiation is numerically addressed with and without control by performing feedforward simulations. The direct control of the radiating surface of the triple panel system is achieved by using a set of d_{33} piezo actuators introducing secondary control signals into the smart panels. Due to the better coherence with the radiated sound field, sound pressure sensors on the radiating side of the vibrating structure are used as error signals.

For the primary acoustic excitation, different incident sound fields are considered. First, the low-frequency noise in a turboprop aircraft cabin is modelled by a sinusoidal signal with frequency of 124 Hz. This frequency simulates the harmonic acoustic excitation due to a three-bladed propeller operating at about 2500 rpm. A multi-channel active noise control system is used to attenuate the sound field radiated by the vibrating panel. Moreover, flat random noise is added to the harmonic noise disturbance in order to achieve a predetermined signal-to-noise ratio representative of a more realistic noise level. Noise attenuation is also evaluated when driving the harmonic disturbance at 197 Hz. Second, the system is excited by a random broadband excitation rather than single harmonic frequencies. Finally, the adaptive feedforward control is tested against transient noise signals varying between 80 and 140 Hz in order to simulate a slightly varying BPF due to a maneuvering flight.

In order to control band-limited noise excitations, three different adaptive signal processing algorithms are used. They are: MELMS, LMMS and scanning error LMS. They are steepest descent-type algorithms which have been shown to be robust and of low computational complexity. The block diagram of the control loop is illustrated in Fig. 4.14. The simulation models, described by Eqs. (4.48)–(4.51), are graphically illustrated in Fig. 4.15. The simulation structure includes characteristics of propagation paths. It is assumed that the sound radiation is monitored at L local points. One reference signal is considered. The block *Loudspeaker* is a matrix of $1 \times L$ transfer functions from the primary source to each error sensor. The block *Active Window* is a matrix of $L \times M$ transfer functions from the secondary sources to the error sensors. The control law is represented by the block *Controller choice* where an adaptive algorithm can be selected among the three possible alternatives. They generate the actuator signals which are linked to the *Active window* model in order to control the effect of the primary disturbance.

120 CHAPTER 4. ADAPTIVE CONTROL OF SOUND RADIATION

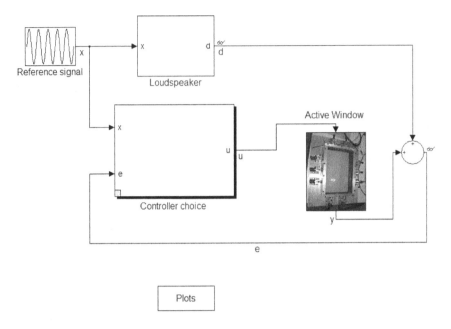

Figure 4.14. Block diagram of the multi-channel control algorithms

The control algorithms are tested at different error sensor positions. As the laboratory test arrangement for the active window differs from an infinite, rigid baffle, the practical application of the actuator concept is considered. The efficiency of the active control system is evaluated by the noise attenuation performance as well as by the required actuation effort. The noise reduction in the area close to the triple-panel system is investigated by placing the error sensors at a distance of 100 mm, 150 mm, 700 mm and 1050 mm respectively from the surface of the radiating panel. The first two acoustic sensors have a practical significance, as people have to stay close to the window in aircraft cabin. The remaining error sensors are representative for noise reduction in the far field.

In the following sections the numerical predictions of the three different feedforward control arrangements are described. Clearly, the goal is to reduce the sound radiated by the controlled triple panel system. A sampling frequency of 6.5 kHz is used throughout the simulations. In Chapter 6, one of the best arrangements, determined in the numerical simulations, will then be programmed on a digital signal processing (DSP) board equipped with a floating point DSP processor for real-time experiments.

4.9. NUMERICAL SIMULATIONS

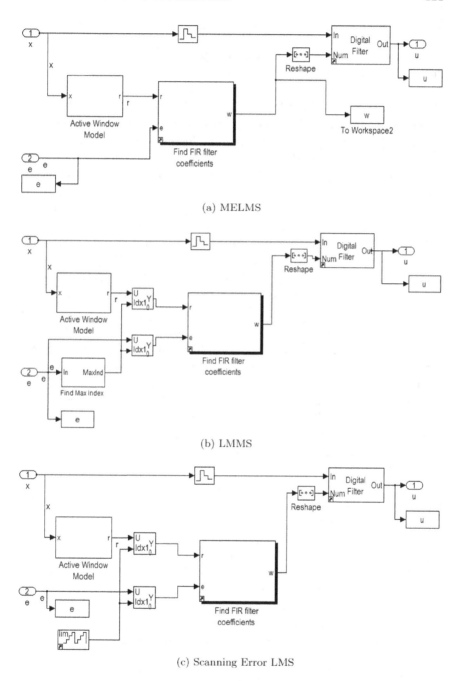

(a) MELMS

(b) LMMS

(c) Scanning Error LMS

Figure 4.15. Adaptive signal processing multi-channel algorithms

4.9.1 *Harmonic primary noise*

In the analyses that follow, preliminary numerical investigations are conducted in the condition where the primary noise source is given by a pure tonal signal. The objective is to find the control signals that must be applied to the piezo actuators in order to minimize the error signals. Such results correspond to the case where the primary acoustic disturbance transmitted through the panels equals approximately 92.5 dB at the error microphone 2.

As a first step, a feedforward control system consisting of four error sensor and one secondary control output is considered. The error signals are used as inputs to the control algorithm while eight actuators are connected in parallel and driven by one channel of the controller. The resulting multi-channel control algorithm is tested by assuming a tonal disturbance at 197 Hz. The noise attenuation results are shown in Fig. 4.16. They are achieved by using a convergence coefficient α equal to 0.003. The curve with square markers represents the noise disturbance computed by knowing the primary transfer functions for each error microphone. The curve with triangular markers represents the anti-noise signal obtained by driving the piezoelectric actuators mounted on the active window prototype. The curve with circular markers is the residual acoustic field achieved with active noise control. The controller gives a perfect cancellation of the sound radiated at the error sensors. It can be observed that for each error microphone, the adaptation automatically converges towards the optimal Wiener filter by adapting the amplitude and phase of the anti-noise signal so that the primary noise wave is completely cancelled. The simulation is performed by using adaptive filter coefficients whose behaviour is shown in Fig. 4.17. The resulting control signals, fed to the active window system, operate in phase opposition with respect to the disturbance signal so as to achieve a global noise attenuation. Starting from zero initial conditions, they rapidly find the optimal values after a small number of iterations.

As discussed in Section 4.8, the primary and secondary paths can slightly vary in time due to changes in environmental conditions, such as temperature. Optimum, Wiener, feedforward controllers, designed using Eq. (4.45), can be very sensitive to small amplitude and phase mismatches due to a time-variant behaviour of the system under control. As shown in Fig. 4.18, even small changes in the optimal phase can largely affect the global control system performance. Furthermore, they produce a greater effect on the sound field than a variation in the amplitude. Adaptive controllers are therefore preferred over the others since they are able to track these changes in the system as well as time-varying disturbances.

4.9. NUMERICAL SIMULATIONS

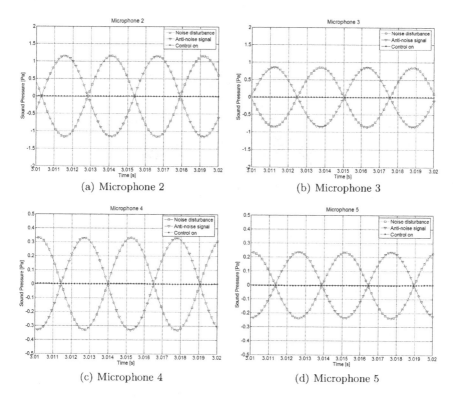

Figure 4.16. Multi-channel control of a tonal disturbance

In order to simulate experimental noise data more effectively, the active noise control performance is also investigated by adding random noise to the microphone signals. As a result, the signals applied to the primary and secondary sources are not perfectly coherent, due to additional uncorrelated noise. This effect aims at replicating the experimental background noise due to spurious sources which may affect the primary noise disturbance. The noise attenuation results are shown in Fig. 4.19. It can be seen that the residual sound field is comparable to the level of random noise added to the primary source.

As a next step, the frequency of the harmonic disturbace is set to 124 Hz. This frequency corresponds to the fundamental BPF of the turboprop aircraft selected in the application scenario. Similarly to the experimental analysis presented in Section 5.5.2, a multi-channel control system based on four secondary sources (piezo 2, 5, 7 and 10) is considered. Results are shown in Fig. 4.20. The radiated sound field is drastically attenuated,

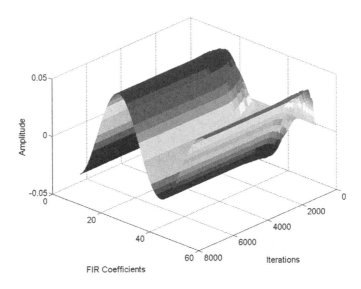

Figure 4.17. FIR filter coefficients obtained for a tonal disturbance

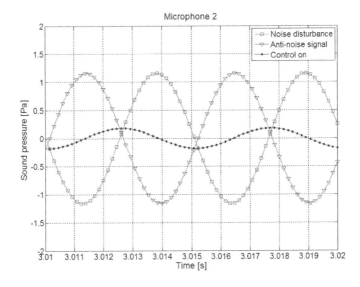

Figure 4.18. Control performance with a variation of 0.05π (rad) in the optimal phase

4.9. NUMERICAL SIMULATIONS

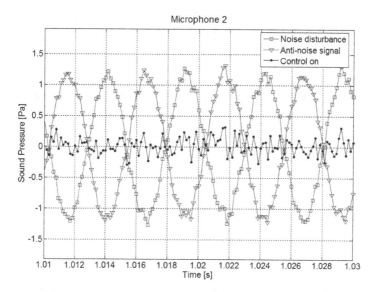

Figure 4.19. Control performance with a random noise added to the primary tonal disturbance

especially at the microphones 2 and 3. The residual acoustic field is of the same order of the random noise added to the processed signals. Due to the higher attenuation in the acoustic transfer function between the primary source and the error sensors 4 and 5, the noise reduction obtained at those microphones is lower. The noise attenuation levels are listed in Table 4.2. A comparison of the root mean square (RMS) pressure of the noise disturbances and that of the error signals is also provided. A maximum reduction of 87.53% is obtained in the controlled noise signal at mic. 2. By converting the RMS pressure signals in decibels of sound pressure level, the noise abatement is equal to 18.08 dB SPL.

4.9.2 Random primary noise

The numerical results described in this section refer to the condition in which the primary source is driven by a band-limited white noise signal in the range 112–562 Hz. Starting with a single channel control, feedforward controllers consisting of multiple secondary actuators and multiple error sensors are investigated. The multi-channel control systems use either one control output driving eight piezoelectric actuators in parallel (piezo 2, 3, 4, 5, 7, 8, 9 and 10), two control outputs driving two secondary sources

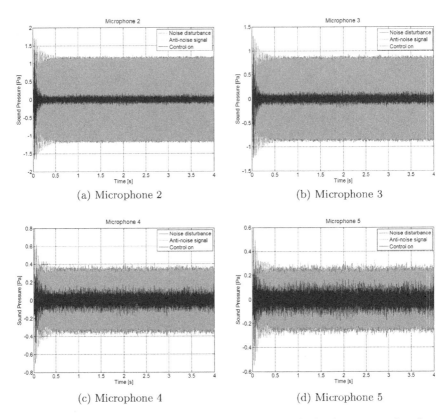

Figure 4.20. Multi-channel control of harmonic primary excitation in presence of random noise

Table 4.2 MELMS control of harmonic noise disturbance: noise attenuation results

	Tonal disturbance at 124 Hz			
	Control OFF (dB SPL)	Control ON (dB SPL)	Signal reduction (%)	Noise reduction (dB SPL)
Control mic. 2	92.27	74.19	87.53	18.08
Control mic. 3	89.61	72.11	86.67	17.5
Control mic. 4	81.47	69.00	76.22	12.47
Control mic. 5	78.59	68.27	69.55	10.32

4.9. NUMERICAL SIMULATIONS

(piezo 2 and 7) or four individual control channels (piezo 2, 5, 7 and 10) to reduce the acoustic pressure radiated in front of the panel. The numbers of output control channels are used for the description of the results. In all three cases, the control performance associated with the minimisation of the pressure at the four error microphones is compared to the case when no control is applied.

In contrast to the results achieved with tonal control, sound attenuation is investigated over a broadband frequency range. The stochastic character of the primary signal makes the active control more difficult with respect to the control of tonal frequencies. The disturbance signal, a random noise input, is fed directly to the controller in order to adapt the filter coefficients. The measured spectra, compared with and without control, yield a rough estimate of the improvements in the insertion loss achievable by the actively controlled structure.

In Fig. 4.21, the results of the adaptive feedforward control are shown by considering a MELMS algorithm with 512 coefficients. The magnitude of the noise radiation detected by the acoustic sensor is reduced significantly throughout the third-octave bands of noise.

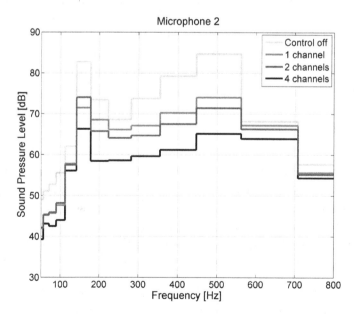

Figure 4.21. Sound radiation with and without the active control engaged when the field pressure is minimized at four error sensors with an increasing number of control actuators

A general attenuation is achieved for all the efficiently radiating vibration modes in the frequency range of interest. Moreover, the sound attenuation tends to improve with an increasing number of controlled secondary actuators. If all eight piezo actuators are driven equally through a single output controller, the sound radiation is reduced by 13 dB around the first resonance. A higher attenuation is observed for higher frequencies up to 300 Hz, if two secondary sources are used to attenuate the response at the error sensors. By extending the number of control actuators to four, the performance is significantly increased. This means that the higher the modal density in the frequency band of interest, the higher the number of actuators required to capture the radiation characteristics of the panel.

In Figs. 4.22–4.24, the time signals obtained at the four error sensors after having implemented the multi-channel controllers described in Section 4.7 are illustrated. For the three LMS-based feedforward simulations, each lasting 200 seconds, the convergence coefficient α has been adjusted until a good convergence speed was achieved during adaptation. The numerical results refer to the condition in which four piezoelectric actuators are driven by using a set of 256 filter coefficients. The trends in the sound radiation demonstrate the amount of random noise which can be actively reduced by using the developed algorithms. Figure 4.25 shows the third-octave band results obtained for the MELMS, LMMS and scanning error LMS algorithms before and after the control system operations. The noise reduction levels are obtained by averaged noise spectra calculated by subdividing the time-domain signals into 200 millisecond blocks. The results achieved for the three cases show meaningful reductions in the band-limited white noise excitation at the sensor positions for all the working frequencies. The profile of the sound pressure level spectra represents a good reference to predict the size of the quiet area obtainable with four control actuators. Although a very similar behaviour is observed in the time domain signals, the overall noise reductions are different. The MELMS and LMMS algorithms succeed in globally reducing acoustic pressure at microphones with higher levels. However, the LMMS algorithm achieves more uniform final levels at the expense of an increase in the mean square values of the individual error signals. Attenuation higher than 20 dB is achieved at some frequencies by using the scanning error LMS algorithm. A very low convergence time, defined as the time at which the squared error falls below ten percent of its initial value, is obtained due to the very small values of α required to keep the algorithms stable.

4.9. NUMERICAL SIMULATIONS

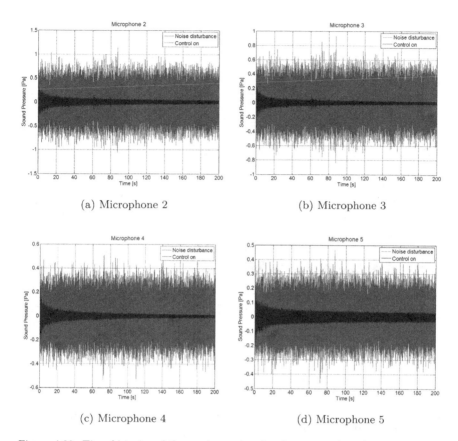

Figure 4.22. Time histories of the random noise signals computed at the error sensors before and after the control system operation. MELMS control algorithm

In all the numerical simulations, the amount of achievable control attenuation is limited by the maximum voltages applicable to the piezoelectric actuators in order to avoid excessive voltage requests which could produce undesirable nonlinear effects. Due to the symmetrical configuration of the actuator elements, similar absolute values of the control voltages have been obtained. For the MELMS simulation, the magnitude of the complex actuation signals driving the actuator 2 is shown in Fig. 4.26. The driving voltage is calculated for a perturbation source strength giving a mean sound pressure level of 117 dB on the incident panel. This excitation level is derived from the band-weighted summation of the levels of the acoustic field at microphone 2, before the controller system operation, by knowing the transmission loss behaviour of the structure at those frequencies. This

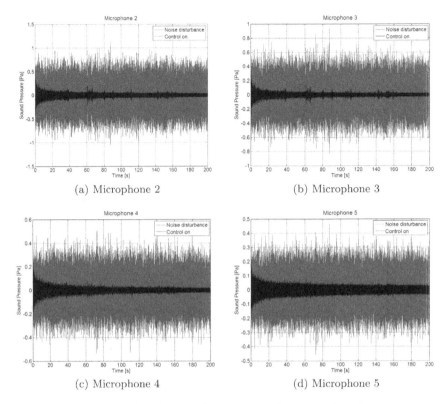

Figure 4.23. Time histories of the random noise signals computed at the error sensors before and after the control system operation. LMMS control algorithm

demonstrates that the control of the radiating panel has the advantage of reducing the control effort with respect to the control of the incident panel. By assuming that these quantities must be multiplied by a factor of 100 due to the voltage amplifier, a more realistic voltage of around 95.84 V(rms) is required to control the structural system and the voltage practical instantaneous limit of ±400 V is not exceeded.

4.9.3 Narrow-band primary noise

In order to illustrate the capability of the developed algorithms to track variations in the primary noise characteristics, a narrow-band disturbance is considered by imposing a bi-directional linear sweep ranging from 80 Hz to 140 Hz and whose frequency changing rate is 30 Hz/s (80 Hz to 140 Hz in 2 seconds, for both increasing and decreasing frequencies). Such an acoustic

4.9. NUMERICAL SIMULATIONS

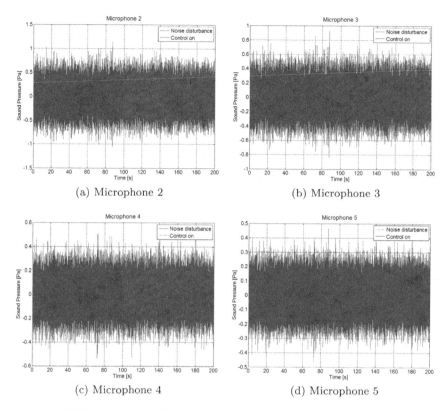

Figure 4.24. Time histories of the random noise signals computed at the error sensors before and after the control system operation. Scanning error LMS algorithm

excitation is chosen to simulate an engine run-up operation before aircraft take-off or the slight variation in the engine's rpm due to maneuvers. The application of an adaptive algorithm, consisting of four error sensors and four control actuators, allows to track these changes by updating the FIR filter coefficients. They evolve in the direction which minimizes the mean-square error using either all error signals, the maximum error signal or only one error signal at each algorithm iteration. The controllers are designed to adjust the amplitude and phase of the reference signal, whose sinusoidal frequency is arranged to be changing in time, in order to drive each secondary actuator properly. The adaptive control is tested by monitoring the behaviour of the filter coefficients with respect to the time-varying disturbance as well as by calculating the global noise attenuation achieved.

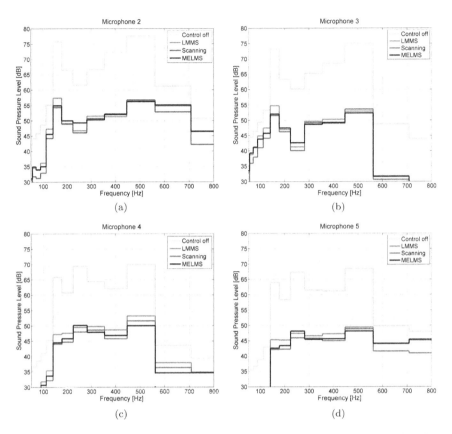

Figure 4.25. Third-octave sound level attenuation achieved in MIMO systems after the ANC system operation in order to cancel random noise using MELMS, LMMS and scanning error algorithms

In contrast to the optimal Wiener filter, which is tuned for only one fixed BPF condition, the adaptive method is fully suited for a sweeping sine wave disturbance where the BPF amplitude and phase change in time. Figure 4.27 shows the numerical attenuations predicted by using the MELMS algorithm. The dark gray signal and the light gray signal are respectively the noise disturbance and the control signal induced by the piezo-stack actuators to the radiating panel. The black curve is the residual acoustic field obtained when the control system is engaged. A convergence coefficient α of 0.0001 is used throughout the simulations as well as 52 adaptive filter coefficients. In Fig. 4.28, the adaptive filter coefficients for the sweeping sine wave disturbance are shown. It can be seen that starting from zero initial

4.9. NUMERICAL SIMULATIONS

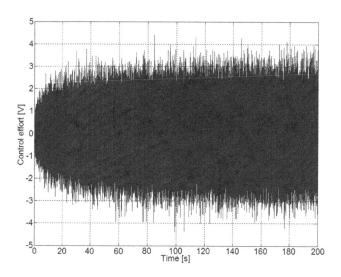

Figure 4.26. Electrical voltage applied to actuator no. 2 for the MELMS control of the random primary noise

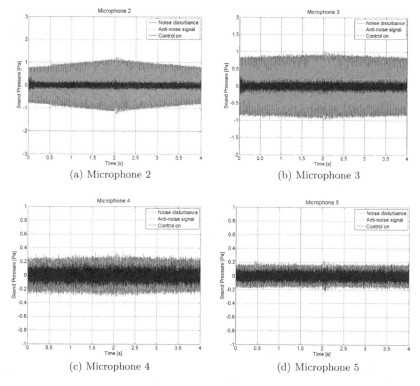

(a) Microphone 2

(b) Microphone 3

(c) Microphone 4

(d) Microphone 5

Figure 4.27. Numerical predictions of noise attenuation for narrow-band disturbance in the range 80–140 Hz

134 CHAPTER 4. ADAPTIVE CONTROL OF SOUND RADIATION

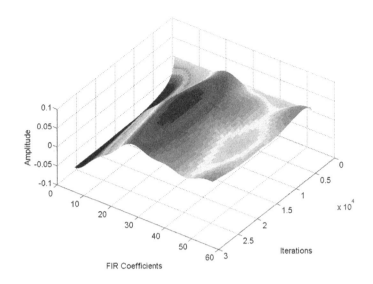

Figure 4.28. FIR filter coefficients for a sweeping sine wave disturbance

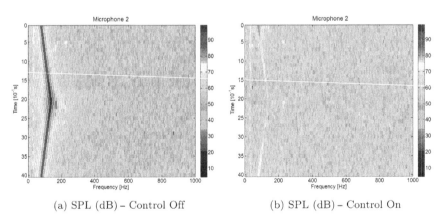

(a) SPL (dB) – Control Off

(b) SPL (dB) – Control On

Figure 4.29. Performance of the MELMS active control algorithm on a narrow-band excitation: (a) spectogram without control, (b) spectogram with control

conditions, they converge rapidly to the optimal values and their magnitude varies with time.

The time-frequency analysis of the primary noise disturbance is illustrated in Fig. 4.29(a). The frequency content of the signal changes between the lower and upper frequency limits of the swept sine. Figure 4.29(b) illustrates the spectrogram plot of the error signal at one of the error sensors

Table 4.3 Noise attenuation [dB] achieved at the error microphones for a sweeping sine wave primary disturbance

	Sweeping sine wave disturbance [80–140] Hz		
	MELMS	LMMS	Scanning
Control mic. 2	18.40	17.52	18.86
Control mic. 3	16.33	15.78	16.27
Control mic. 4	10.80	11.85	10.94
Control mic. 5	4.91	5.65	5.26

with control. Such a spectogram is obtained by the average of the STFT of the signal calculated at regular intervals of time (0.1 seconds). A flat random noise, whose frequency contribution can be observed, is added to the primary signal in order to simulate a more realistic acoustic disturbance. Although the control system performance is limited by the random noise added to the error signals, a perceptible attenuation is achieved for all the error sensors. A mean reduction of 87.9% in the disturbance signal is obtained at the microphone 2, corresponding to a noise attenuation of 18.40 dB in the overall sound pressure level. The reduction always remains important even in the other error sensors.

A comparison between the sound pressure levels obtained with and without control using the three adaptive methods is listed in Table 4.3. No remarkable differences are achieved in the attenuation results. Notice that they are computed as the mean difference between the level of the acoustic field before and after the active control. A faster noise reduction is obtained by the MELMS algorithm as it uses all the error signals. On the other hand, the LMMS algorithm provides a more uniform residual field compared with the MELMS and scanning error LMS. This could be expected since its control strategy is based on the minimization of the maximum error signal. The numerical results obtained for the MELMS algorithm will be validated through MIMO experiments. The experimental results will be then used to predict passenger comfort benefits in aircraft cabin.

4.10 Summary

The most relevant error criteria for adaptive control of sound radiation involve the total radiated sound power, under free field conditions, or the acoustic potential energy, for enclosed sound fields. Unfortunately, these quantities are difficult to be measured in a practical adaptive controller, but they can be reasonably well approximated by considering the sum of the squares of a number of error microphones positioned on the radiated field. The cancellation of acoustic pressure at fixed points by active noise control yields a quadratic cost function which can be written in a general Hermitian quadratic form.

In order to cope with time varying disturbances, the development of a feedforward controller modified so as to adapt itself to a changing primary field has been presented. The steepest descent algorithm based on the instantaneous gradient has been introduced, as obtained from a measurement at each time instant, instead of the exact gradient requiring the knowledge of the cross spectral matrices.

The effectiveness of a multi-channel ASAC controller in suppressing the sound radiation from a benchmark test-bed has been numerically investigated by using experimental plant models. It has been assumed that the dynamic behaviour of the system to be controlled did not change during control, and could be measured beforehand. Control structural inputs to the piezoelectric actuators have been computed so as to reduce the radiated sound field. The signals driving the secondary actuators have been obtained by properly filtering the reference signal via a digital controller. An accurate knowledge of the transfer functions between the actuators and sensors has been achieved by experimental system identification. Due to the adaptation of the control filter, the resulting feedforward algorithm has not been open-loop.

The control system performance has been evaluated for tonal frequencies, narrow-band and broadband noise. Theoretical FIR filter coefficients able to suppress harmonic disturbances at 124 and 197 Hz as well as narrow-band primary noise between 80 and 140 Hz have been calculated with and without adding random background noise to the imposed disturbance. A band-limited white noise excitation has been also considered in the range 112–562 Hz. The derivation of three MIMO controllers, namely the MELMS, the LMMS and the scanning error LMS algorithms, has been discussed for use in numerical simulation studies. Noise attenuation achieved at four selected field points, located at different distances from the radiating panel, has been investigated by increasing the number of actuators used for control. The adaptivity of the self-tuning noise controller has been also

4.10. SUMMARY

evaluated for a frequency sweeping acoustic excitation by comparing the spectograms obtained with and without the active control system engaged. Noise simulation results have demonstrated that the insertion loss of the triple-panel partition can be significantly improved by controlling the sound field monitored by external error sensors.

Chapter 5

Noise-Reducing Smart Windows

5.1 Introduction

Identifying the operational source strength for each noise source and the dominant transfer paths to the cabin interior noise is crucial to attenuate vibration and noise levels through active noise control. However, a correct interpretation of the individual contributions of multiple paths is not straightforward, especially at low and mid frequencies. Sound transmission through aircraft window panels is an important area of research that has attracted the attention of this book. This results from a number of studies assessing the impact of aircraft windows on the total sound transmission loss behaviour of fuselage sidewalls. As Fig. 5.1 shows, in order to realize an acoustically efficient sidewall design, the window and the surrounding wall area shall have approximately the same sound transmission loss properties [54].

Passenger cabin windows thus play an important role in the acoustic performance of the whole aircraft fuselage structure. Nevertheless, although the acoustic behaviour of double pane aircraft-type windows has been extensively investigated in the literature [118–120], only a few papers provide numerical and experimental data for triple pane windows. Multi-glazed windows provide good sound insulation only at relatively high frequencies. Due to the mass–air–mass resonance, their acoustic performance rapidly deteriorates at low frequencies, where the sound transmission loss can even become worse than that of a single partition. This resonance frequency depends on the mass and the distance of the panels as well as the stiffness of the cavity medium.

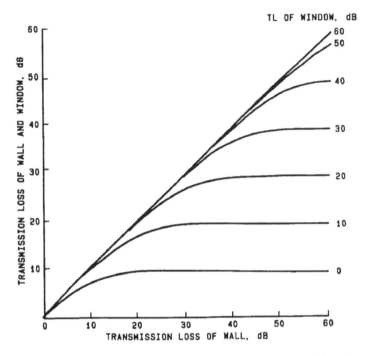

Figure 5.1. Estimated transmission loss of window and aircraft sidewall (window area = 10% of total area) [54]

In the low frequency domain, aircraft windows exhibit poor sound transmission loss capabilities and hence a significant amount of acoustic energy is potentially radiated into the cabin interior. For a propeller driven aircraft, such a behaviour results in higher interior noise levels, since the highest acoustic excitation typically occurs in the frequency region between 70 and 180 Hz [118]. It then seems natural to investigate the possibility of improving low-frequency sound insulation characteristics of multi-wall windows by means of active control techniques.

Since most of the passive means to increase transmission loss around the mass–air–mass resonance requires additional volume and weight, the use of smart systems capable of adaptively enhancing the low frequency transmission loss of noise into the cabin is thus an important target for the next generation aircraft design. However, the integration of smart actuators and sensors into transparent panels remains a challenging problem since, in principle, no equipment should interfere with the view through the window transparency. Thus, although improving transmission loss of actively

5.1. INTRODUCTION

controlled multi-glazed windows seems very promising, the effectiveness of active solutions still has to be proved.

One of the purposes of this book is to validate experimental applications of active control of sound transmission through smart panels in order to enhance their own sound insulation characteristics. This objective requires the exploitation of light-weight piezoelectric actuators and their integration into an application-level smart solution as well as the development of a control algorithm and its implementation on a real-time controller. The benchmark study, which is extensively investigated throughout the body of this book, involves the design of an aircraft-type window test-bed capable of reducing sound transmission into aircraft cabins. The application scenario regards the interior noise control of a turboprop aircraft. This is because at low frequencies the cabin of a propeller-driven aircraft is more annoying than that of a turbofan aircraft, due to the low-frequency discrete tones related to the blade passage frequency (BPF). Such an adaptive structure is proposed and potentially evaluated as a replacement for the currently used ANC systems.

The low-frequency noise cancellation capabilities are numerically assessed and experimentally validated. The control strategy relies on a feed-forward control scheme and hence on the availability of reliable prior information about the unwanted noise. The reference signal is derived from the BPF of the propeller, sensed by a tachometer, to modify the interior acoustic field through feedback of acoustic sensors distributed throughout the fuselage cavity. Contrary to the ANC approach shown in Fig. 5.2, aiming at directly attenuating the sound field in the cabin by means of acoustic speakers, interior noise control is achieved by reducing sound transmission into the cabin passenger area by means of an adaptive structure driven by vibration actuators. Figure 5.3 shows a sketch of the smart window concept installed onboard the aircraft. Following a passenger-focused approach, the control noise field is measured by microphones mounted on the headrest of the seat place inside the cabin.

The work presented in this chapter follows an application-oriented strategy for designing an actively controlled aircraft-type window demonstrator based on panel control. A triple-wall configuration is chosen to enhance the passive sound insulation properties predicted by means of the developed sound transmission simulation tool. Furthermore, smart actuators are exploited for structure-acoustic control of the radiating panel of the triple-panel system.

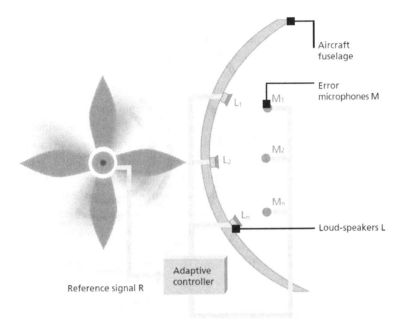

Figure 5.2. Active reduction of interior noise in turboprop aircraft by ANC [121]

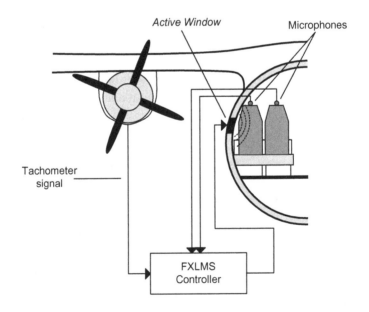

Figure 5.3. The smart window concept applied to a turboprop aircraft

The developed structure is numerically modelled first. The *in vacuo* panel's natural frequencies are computed as well as the coupled vibro-acoustic response of the triple-panel partition, including the modelling of the fluid between the panes. A surface-based contact approach is used to couple the structure with the inner air. Then, the experimental prototype is presented. An experimental modal analysis is carried out in order to validate the FE predictions. Due to the important role in the radiation properties of the triple-wall structure, particular attention is paid to the dynamic characterization of the radiating panel. The mode shapes, eigenfrequencies and the frequency response functions of the coupled triple-panel system are validated experimentally. Then, the validated modal model is employed to estimate the sound transmission loss behaviour of the triple-wall structure by using the developed FEM/Rayleigh approach. Finally, the smart actuator configuration is experimentally investigated by evaluating the effectiveness in exciting the coupled modes to be controlled.

5.2 Literature Survey

A brief literature review on active and passive control of sound transmission through multi-wall partitions is reported in the following. This overview focuses mainly on the different control strategies available for aircraft applications. Panel control and cavity control are discussed in two separate sections.

The design of an acoustically-optimized aircraft window for interior noise control is presented in [119]. An elliptical window shape, having the major axis about twice the length of the minor axis, is capable of meeting two competing design targets: the minimization of any sharp corners in the window that cause high stress levels affecting the structural integrity of the aircraft, and the maximization of the ratio of window transparent area to the amount of acoustic noise transmitted through the window.

Sound transmission through windows can be reduced in different ways. In general, noise insulation properties increase proportionally with the window mass. Increasing window thickness is an obvious way of increasing mass but, in some cases, this may be impractical due to the bulk and weight constraints. Laminated glass, consisting of two or more glass sheets with one or more interlayers of transparent viscoelastic damping material sandwiched between the layers, is a widely used solution for automotive and architectural applications [122–125]. This provides better acoustic performance than monolithic glass and it is inherently safer in the event of breakage. The plastic interlayers between the glass panes determine significant benefits in the sound transmission loss at the medium-high frequencies. This also minimizes the possibility of glass fragments injuring passengers during a collision.

While laminated glass has shown potential in noise control applications, little work has been performed on plexiglass. Multiple layers of plexiglass windows have been tested at NASA Langley Research Centre by assessing the transmission loss for a diffuse acoustic excitation as well as the radiated sound power for a point force excitation [126–129]. Results have demonstrated that single-panel windows have limited performance in applications requiring high transmission loss capabilities. Doubling or tripling the thickness of single panel windows provides only marginal improvements in terms of transmission loss, while adding significant weight and cost.

The airborne sound insulation of windows can be increased by dividing the element into two or more glass panes separated by air gaps. By taking advantage of the spacing between the window panes, multi-pane window assemblies provide better noise insulation than an equivalent single-pane window, especially at medium and high frequencies. However, at some frequencies, the transmission loss can correspond to or can be even less than that of a single panel of equal weight. This is because the air chamber between the panels acts like a spring transferring the vibrating energy from one panel to the other. The resonance frequency at which the sound transmission loss is minimum is usually referred to as mass–air–mass frequency.

Experimental investigations on the sound reduction capabilities of double- and triple-pane windows with various thicknesses and distances between the glasses can be found in [95, 130–133]. Passive sound transmission through simply supported double-panel partitions with enclosed air cavity has been investigated by a number of researchers. Some examples are given in [134–136]. A number of common results emerge from these experiments. By applying additional soundproofing materials, double- and triple-glazed windows provide enhanced sound transmission loss above 1000 Hz. In the low frequency domain, passive sound insulation methods are less effective. The major drop in the sound insulation properties occurs around the mass–air–mass resonance frequency where the cavity pressure is dominated by the first cavity mode which is strongly coupled with the window plates.

Vaicaitis [137] conducted noise transmission experiments into the cabin of a turboprop aircraft in order to evaluate the noise performance of a variety

5.2. LITERATURE SURVEY

of sidewall acoustic treatments designed for interior noise control. Moreover, additional tests were performed in order to quantify the amount of noise transmitted through double-wall aircraft windows. Results showed that the motions of the exterior and interior panels of windows are strongly coupled with the air cavity that separates them. For frequencies up to 250 Hz, the acceleration levels in the interior plexiglass panel were higher than those in the exterior panel of the double-pane window. A double wall resonance was observed at about 330 Hz which increased the level of transmitted noise. Similar experiments were conducted by Hayden *et al.* [53].

In [54], Prydz *et al.* present a theoretical analysis of multi-pane window assemblies for advanced turboprop aircraft. Over 200 configurations are analyzed to meet specific noise requirements in terms of noise transmission properties at the estimated blade passage frequency. Thickness and materials of window panes as well as the spacing between the partitions are optimized with respect to the sound attenuation capabilities of the window assembly. The optimal system design consists of two glass outer panes separated by a small air space and a third fail-safe inner panel made of plexiglass. An experimental verification of the predicted transmission loss behaviour is also presented.

Buehrle *et al.* [126, 127, 129] studied single-wall laminated windows for aircraft interior noise control. They consisted of two or three layers of plexiglass with transparent viscoelastic damping material sandwiched between the layers. For damped laminated windows, nominally 6.35 mm thick, an increase in the sound transmission loss of 4.5 dB was experimentally achieved over a frequency range between 50 Hz and 4 kHz. Furthermore, an enhanced transmission loss behaviour was obtained for window assemblies consisting of two laminated glass panels separated by an air gap.

In [138], the acoustic transmissibility of aircraft triple pane windows is experimentally investigated. They are designed to provide 69 dB noise transmission loss at the blade propeller frequency (164 Hz) of a turboprop aircraft. The experimental campaign was carried out at the NASA Langley Research Center. A comparison with the theoretical predictions detailed in [54] is also discussed. Double- and triple-wall resonances are found to degrade the transmission loss characteristics within the 200–500 Hz one-third octave bands.

Active control is a suitable alternative for improving the low-frequency sound transmission loss performance of aircraft windows. Active control of flat homogeneous multi-wall structures has been addressed in several scientific works, mainly focused on automotive and architectural applications, such as car shells, noise blocking partition walls, etc. Three different approaches can be found in the literature: (i) panel control, (ii) cavity control and (iii) room control. The first class of methods in this field deals with active structural acoustic control by means of vibration sources mounted on either of the panels so as to reduce the radiated sound field [87, 139]. The second method focuses on active control of the cavity sound field between the panels through the use of secondary sound sources [140–143]. These acoustic sources allow pumping air in and out of the cavity in an attempt to block airborne sound transmission. Both methods require the use of either structural sensors, such as piezoelectric sensors and accelerometers, glued or embedded in the radiating panel or pressure sensors, such as microphones, measuring either the radiated sound field or the sound field within the cavity. The third method, usually referred to as room control, aims at controlling the acoustic field in the receiving room rather than the sound transmission through the structure [116]. This is achieved by means of counteracting acoustic sources.

5.2.1 *Control of sound transmission through multi-wall partitions: panel control*

One of the first published contributions on active control of acoustic-structural coupled systems was provided by Deffayet *et al.* [144]. They aimed at reducing the acoustic radiation of a baffled rectangular plate by using secondary acoustic monopole sources. Nevertheless, the use of vibration sources as control actuators fixed at a radiating panel in such a way as to minimize the acoustic radiation into the receiving space was first introduced by Fuller *et al.* [44, 145]. They coined the term ASAC (active structural acoustic control).

A considerable amount of research involving ASAC of multi-wall structures deals with homogeneous flat panel systems. By implementing the basic principles of ASAC, the noise transmission through multi-wall structures can be controlled by feeding back structural or acoustic error signals to structural actuators. A cost function related to the observed far field pressure can be minimized by using a number of sensor and actuator pairs. This does not necessarily imply a suppression of the vibration amplitudes of the structure since the vibration may be, in principle, even increased to reduce the sound radiation from the structure. This mechanism is usually referred to as modal restructuring and involves forcing the panel to become a less

5.2. LITERATURE SURVEY

efficient acoustic radiator by means of destructive interferences between the contributing radiation modes. Vibration actuators and sensors, such as piezoelectric devices, can be bonded or embedded into multi-wall structures to modify vibrations, and hence the way the structure radiates sound, through their electro-mechanical coupling properties. On the other hand, since each vibration mode has different radiation efficiency and some modes can be better coupled to the resulting acoustic field than others, only the few dominant radiation modes require controlling. This method is called modal suppression. It aims at attenuating those structural modes which show high radiation efficiencies, thus reducing the structural sound radiation as well as the mean vibrational level in the structure.

In [139], Carneal *et al.* present some comparative studies on the active control of the incident and the radiating panel of a double-wall system. The actuation is made via a set of piezoelectric patches aimed at controlling the sound radiation resulting from a harmonic excitation. A set of microphones placed in the receiving chamber provides the error signal for a feedforward controller. The authors conclude that, due to the direct effect on the radiated sound field, controlling the vibration of the radiating panel in a way that minimizes sound radiation results in a better performance in terms of transmission loss than controlling the vibration of the first sound-receiving panel, i.e. the incident panel of the double wall structure. In addition, the control of radiating panels excited in resonance condition is more effective than off-resonance active control. In [100], the authors perform further analytical and experimental investigations to validate such results. A comparison of the theoretical and experimental uncontrolled transmission loss is provided and the potential for applying ASAC to reduce the interior noise of an aircraft is numerically addressed. Design guidelines for active control of double panel partitions by ASAC are also provided.

Bein *et al.* [146, 148, 149] present an actively controlled lightweight facade specimen consisting of a double-glazed window. The ASAC system consists of distributed sensors and laminar piezoceramic actuators placed at the margin of the laminar elements of the facade, so that they can be hidden behind the plates without covering the view. The actuator positioning of the laminar elements on the surface is optimized on the base of a controllability criterion derived by modal strains in the planar directions of the plates. The use of collocated SISO adaptive controllers results in a noise reduction up to 10 dB. The radiated acoustic power is reduced by 7 dB.

In [147], Sagers *et al.* present a promising approach involving active segmented partitions. An experimental analysis is performed on a double-panel system in which partitions are segmented and independently controlled by an analog feedback controller. Transmission loss results show an improvement of about 15 dB over the band from 20 Hz to 1 kHz.

In [150–152], Kurtze et al. present a prototype of active noise absorbing double-glazed window. Distributed laminar piezoceramic actuators are used to induce vibrations into the plate-like elements that counteract the structural vibrations caused by airborne path, thus reducing the radiated noise. In addition, piezoceramic stack actuators are integrated into the frame to create an adaptive interface preventing structure-borne sound from being transmitted into the interior. Structural responses to disturbances such as white and pink noise as well as pure sine signals are investigated and a reduction of up to 10 dB is experimentally achieved.

In [153, 154], transparent actuators, consisting of a piezoelectric poly (vinylidene fluoride) (PVDF) thin film coated with compliant carbon nanotube-based transparent conductors are designed and manufactured for the active control of sound transmission through windows. The developed thin-film speaker has the advantages of being flexible, transparent, thin and lightweight. These actuators enable the panel to vibrate as a distributed source, thus differing from conventional loudspeakers which operate in piston mode. However, such actuators show significant limits in the sound power output at low frequencies.

In [155–158], Zhu et al. investigate thin glass panels acting as transmission preventers for the propagation of sound. An adaptive feedforward control system is developed to drive the transmitted sound to zero. The glass panels are able to effectively block transmission of sound by 20 dB in the case of tonal frequencies and by 10–15 dB in the case of broadband noise [155].

5.2.2 Control of sound transmission through multi-wall partitions: cavity control

As mentioned earlier, the low frequency behaviour of multi-wall partitions is affected by the mass–air–mass resonance phenomenon. A suitable approach for enhancing the transmission loss of multi-wall partitions is the active control of the cavity resonances. In [159], cavity control is used to reduce sound transmission through a transparent double panel partition by using both acoustic and piezoceramic film actuators. A comparison between conventional loudspeakers and distributed acoustic actuators, based on active PVDF material integrated in the boundaries of the cavity, is provided. Increased acoustic energy in the cavity has been noticed at frequencies where

5.2. LITERATURE SURVEY

the double panel radiates much sound energy. Reducing acoustic potential energy in the cavity has been found as effective as controlling the structural vibration energy or the radiated sound field. Meaningful reductions in the sound transmission loss have been achieved below 300 Hz.

Jacob et al. [140, 141] experimentally investigated the cavity control of double-glazing windows by means of loudspeakers and microphones inside the air cavity between the panels. In [140], an adaptive feedforward controller is developed and tested by using different combinations of acoustic actuators grouped together or driven individually. Improvements of more than 12 dB are obtained in the frequency region of the mass–air–mass resonance with an optimized arrangement of loudspeakers suppressing the first cavity mode. The reference signal is taken directly from the primary signal. In [141], an adaptive feedback controller is realized via an internal module reconstructing the reference signal from the knowledge of the error signals and the secondary paths. Only 8 dB of reduction is achieved due to the higher computational effort necessary to derive the reference signal estimate for the adaptive feedback controller.

In [160] numerical and experimental studies addressing active control of sound transmission through a double-wall fuselage structure are presented. Due to the high modal density of the partition, multiple speakers are used at optimized locations. Experiments demonstrate a correlation of the acoustic potential energy in the cavity with the radiated sound power. In [161], an adaptive feedforward LMS controller is implemented to reduce the transmission of sound through double-glazed windows by driving speakers in the cavity between the panes. A FE-BE model of the vibro-acoustic system is also built. For a system with two speakers and two microphones, experiments show a noise reduction between 5 and 10 dB in the frequency range between 30 and 160 Hz.

A comparison between the panel and cavity control is carried out in [162–164]. In [164], the control of airborne and structure-borne sound through an aircraft sidewall is performed by active mounts connecting the trim and the skin panels and by loudspeakers between the panels. The former are driven to cancel the out-of plane velocities of the trim panel by acting in a direction normal to the panels. However, due to the structural coupling occurring via the frame-mounting system, a significant fraction of energy is transmitted via the peripheral paths of the frame. The control loudspeakers are placed near the corners of the interior cavity to control the coupled structural acoustic modes. The total sound power radiated, the structural velocities of the radiating panel and the acoustic pressure between the panels are respectively minimized to compare the control performance. The numerical results show that cavity control provides higher attenuation of sound. No significant reductions are achieved with active mounts placed at the corners of the two panels except when used in combination with the control loudspeakers. Due to the low control authority achieved by positioning the structural actuators at the ends of the panel, neither the mass–air–mass sound transmission mechanism nor the volumetric vibration of the two panels are effectively controlled.

Similar conclusions can be drawn from the studies in [162, 163]. Contrary to the results achieved by Carneal et al. [100, 139], the authors conclude that for cavity-dominated modes, cavity control performs better than panel control. The use of secondary loudspeakers and error microphones inside the window air-cavity is more effective than controlling the vibration of one of the plates by vibration inputs. This behaviour can be explained by the lower acoustic modal density [165]. The first cavity mode, i.e. the (0,0,0) mode, which dominates the pressure in the cavity at the mass–air–mass resonance, is much easier to control with a limited number of acoustic actuators and sensors rather than vibration inputs. Furthermore, away from the mass–air–mass resonance frequency, the sound field in the cavity exhibits a simpler shape than the structural response.

Nevertheless, it is crucial for the success of cavity control that the secondary sound sources are effectively integrated in "aircraft-like" double-panel partitions. However, although loudspeakers are currently cheap and widely used components, the readily available commercial off-the-shelf products do not meet the required characteristics of being both compact and powerful. Due to their size and weight, classical voice coil loudspeakers cannot be inserted in the narrow double panel partitions of aircraft fuselages or windows. This is clearly demonstrated by the unusually large interpanel spacing adopted in many scientific works dealing with cavity control in order to simplify the mounting of the control loudspeakers [101]. Novel concepts such as flat loudspeakers are not yet capable of producing large sound pressures without significant distortions in the low frequency range, where passive treatments are insufficient and substantial acoustic benefits would be necessary. Additionally, the costs associated with their development and manufacturing are still excessive in comparison with standard passive insulation products.

As a result, although both approaches are in some sense equivalent [101], active control of sound power radiated by the outer plate of a multi-wall

5.3. PIEZOELECTRICITY

structure mantains practical advantages with respect to cavity control [100]. For the purpose of controlling aircraft interior noise field, the direct sound radiation from the vibrating side walls can be efficiently controlled by configuring the actuators as a distributed acoustic boundary.

5.3 Piezoelectricity

Piezoelectricity is a link between electrostatics and mechanics. Piezoelectric materials have the ability to convert mechanical energy into electrical energy and vice versa. When a piezoelectric crystal is deformed in a given direction, this produces an electric voltage proportional to the applied stress. The resulting electrical charge is proportional to the externally applied force. This effect, which is shown in Fig. 5.4, is called the direct piezoelectric effect. The piezoelectric effect is a reversible effect. When subjected to an externally applied voltage, the piezoelectric crystal deforms and the magnitude of the developed strain depends on the field strength. This effect is called the inverse piezoelectric effect.

The special ability in converting electrical into mechanical energy and vice versa, for static and dynamic applications, makes piezoelectric materials particularly effective in the shape and vibro-acoustic control of adaptive structures. A more general overview of multi-functional materials used for designing smart structures is given in [167, 168]. The electrical and mechanical coupled constitutive equations of piezoelectric materials are widely known. By using the notations of the IEEE standard on piezoelectricity

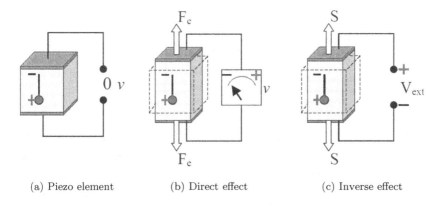

(a) Piezo element   (b) Direct effect   (c) Inverse effect

Figure 5.4. Piezoelectric effects [166]

[169], they read:

$$\{T\} = [c^E]\{S\} - [e]^T\{E\} \tag{5.1}$$
$$\{D\} = [e]\{S\} + [\epsilon^S]\{E\}. \tag{5.2}$$

In Eq. (5.1), the piezoelectric coupling coefficients matrix $[e]$ relates the stress $\{T\}$ to the electric field $\{E\}$ in the absence of mechanical strain. The matrix $[c]^E$ is the elasticity matrix for a constant electric field. In Eq. (5.2), $[e]$ relates the electric charge per unit area $\{D\}$ to the strain $\{S\}$ under a zero electric field (short-circuited electrodes). Piezoceramic manufacturers usually provide the strain-charge form $[d]$ of the 3 × 6 piezoelectric coupling matrix, derived from $[e]^T = [c]^E [d]^T$. Finally, $[\epsilon^S]$ is the permittivity under constant strain.

The general equations for a piezoelectric actuator (inverse piezoelectric effect) and a piezoelectric sensor (direct piezoelectric effect) can thus be written as [169]:

Actuation:

$$\begin{Bmatrix} S_{11} \\ S_{22} \\ S_{33} \\ 2S_{23} \\ 2S_{13} \\ 2S_{12} \end{Bmatrix} = \underbrace{\begin{bmatrix} s_{11} & s_{12} & s_{13} & 0 & 0 & 0 \\ s_{12} & s_{22} & s_{23} & 0 & 0 & 0 \\ s_{13} & s_{23} & s_{33} & 0 & 0 & 0 \\ 0 & 0 & 0 & s_{44} & 0 & 0 \\ 0 & 0 & 0 & 0 & s_{55} & 0 \\ 0 & 0 & 0 & 0 & 0 & s_{66} \end{bmatrix}}_{\text{compliance}} \begin{Bmatrix} T_{11} \\ T_{22} \\ T_{33} \\ T_{23} \\ T_{13} \\ T_{12} \end{Bmatrix} + \underbrace{\begin{bmatrix} 0 & 0 & d_{31} \\ 0 & 0 & d_{32} \\ 0 & 0 & d_{33} \\ 0 & d_{24} & 0 \\ d_{15} & 0 & 0 \\ 0 & 0 & 0 \end{bmatrix}}_{\text{coupling}} \begin{Bmatrix} E_1 \\ E_2 \\ E_3 \end{Bmatrix} \tag{5.3}$$

Sensing:

$$\begin{Bmatrix} D_1 \\ D_2 \\ D_3 \end{Bmatrix} = \underbrace{\begin{bmatrix} 0 & 0 & 0 & 0 & d_{15} & 0 \\ 0 & 0 & 0 & d_{24} & 0 & 0 \\ d_{31} & d_{32} & d_{33} & 0 & 0 & 0 \end{bmatrix}}_{\text{coupling}} \begin{Bmatrix} T_{11} \\ T_{22} \\ T_{33} \\ T_{23} \\ T_{13} \\ T_{12} \end{Bmatrix} + \underbrace{\begin{bmatrix} \epsilon_{11} & 0 & 0 \\ 0 & \epsilon_{22} & 0 \\ 0 & 0 & \epsilon_{33} \end{bmatrix}}_{\text{permittivity}} \begin{Bmatrix} E_1 \\ E_2 \\ E_3 \end{Bmatrix}. \tag{5.4}$$

With the assumption of poling along the third axis, when an electric field E_3 is applied, an extension along the same direction is produced. This effect is governed by the piezoelectric coefficient d_{33}. At the same time, a shrinkage occurs along the directions 1 and 2, perpendicular to the electric field, whose amplitude is controlled by d_{31} and d_{32}, respectively. Equation (5.3) also indicates that an electric field E_1, normal to the direction of polarization

5.3. PIEZOELECTRICITY

Table 5.1 Typical properties of piezoelectric materials [106]

Material properties	PZT	PVDF
Piezoelectric constants		
d_{33} (10^{-12} C/N or m/V)	300	-25
d_{31} (10^{-12} C/N or m/V)	-150	uni-axial
		$d_{31} = 15$
		$d_{32} = 3$
		bi-axial
		$d_{31} = d_{32} = 3$
d_{15} (10^{-12} C/N or m/V)	500	
$e_{31} = d_{31}/s^E$ (C/m^2)	-7.5	0.025
Dielectric constant ϵ^T/ϵ_0	1800	10
$\epsilon_0 = 8.85 \ 10^{-12}$ F/m		
Young's modulus $1/s^E$ (GPa)	50	2.5
Max. stress (MPa)		
Traction	80	200
Compression	600	200
Max. strain	Brittle	50%
Max. operating T (°C)	80°–150°	90°
Coupling factor k_{33}	0.7	0.1
Max. electric field (V/mm)	2000	$5 \ 10^5$
Density (kg/m^3)	7600	1800

3, produces a shear deformation S_{13} controlled by the piezoelectric constant d_{15}. Similarly, S_{23} is generated by a field E_2 due to the piezoelectric coefficient d_{24}.

The most widely used piezo materials for actuation and sensing are lead-zirconate–titanate ceramics (piezoceramics or PZT) and polyvinylidene fluoride polymer (PVDF). Some exemplary properties are given in Table 5.1. Notice that the piezoelectric coefficient e_{31} is 300 times larger for PZT than for PVDF. This suggests that piezoceramics exhibit higher actuation capability than piezopolymers. For this reason PVDF films are mainly used as sensors in active noise and vibration control. Figure 5.5 shows both the linear and laminar piezoelectric actuator/sensor configurations. Some preliminary numerical investigations involving laminar piezotransducers are presented in Appendix A.

Stack actuators, centre of interest for the applications detailed in this book, consist of thin ceramic disks separated by electrodes. The electro-mechanical coupling properties, describing the conversion of the electric energy in mechanical energy and *vice versa*, are such that the piezoelectric constant d_{33} dominates the other coefficients in the constitutive equations. As a result, an electric field parallel to the direction of polarization induces an

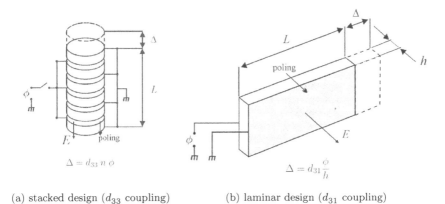

(a) stacked design (d_{33} coupling)   (b) laminar design (d_{31} coupling)

Figure 5.5. Piezoelectric actuation/sensing design [106]

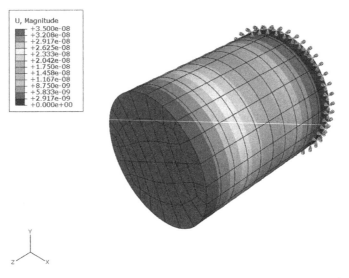

Figure 5.6. Contraction of a clamped-free piezoelectric stack actuator subjected to a DC voltage of 100 V

expansion of the ceramic. Unlike the laminar design where the useful direction of expansion is normal to the electric field, the polarization axis and the applied electric field of the piezoelectric stack coincide with the direction of expansion of the linear actuator. Figure 5.6 shows the induced displacement of a clamped-free piezoelectric stack actuator ($d_{33} = 3.5 \ 10^{-10}$ C/N

5.3. PIEZOELECTRICITY

or m/V) driven by a DC voltage of 100 V. Such simulation is performed in ABAQUS®. The equipotential condition for the electrode surfaces is ensured by the *EQUATION option linking the surface electrical potential to the electrical potential of a master node.

Stacked piezoceramic actuators provide highly attractive features with respect to piezoelectric patches for the active control of sound radiation from vibrating structures. The main advantages are the high bandwidth and the cability of imparting relatively high actuation force, which can be employed for the structural control of low wave number coupled modes [65, 170, 171]. The major limits, however, are related to the low stroke capability. Since the maximum expansion in tension can only reach values in the order of 0.1% for typical PZT ceramics, high driving voltages may be required to realize effective active control systems. For this reason, stacked discs are usually paired, reducing the required electric field within each layer. A compression preload is also necessary in order to compensate the low tensile strength.

When no external load is applied, the maximum actuator displacement is related to the maximum voltage applied by the approximate relationship

$$\Delta l_{max} = d_{33} \cdot n \cdot \phi_{max}. \tag{5.5}$$

If the actuator is mounted against a supporting host structure with an infinite stiffness, Δl becomes zero and the maximum force is exerted on the host structure. This force is denotes as blocked force F_B

$$F_B = k_p \cdot d_{33} \cdot n \cdot \phi_{max} = k_p \cdot \Delta l_{max} \tag{5.6}$$

where k_p is the actuator stiffness defined as $k_p = E_a S_a / l_a$. In this case, the piezo acts as a force generator.

In practical applications, the effective force, F_{eff}, generated by a piezo actuator acting against an external spring load of stiffness k_s is

$$F_{\text{eff}} = k_p \cdot \Delta l_{max} \left(1 - \frac{k_p}{k_p + k_s}\right). \tag{5.7}$$

An example of the actuator characteristic curve is given in Fig. 5.7 for different driving voltages. It follows the line passing through the blocking force and the free displacement points described above. The intersection between the stiffness of the host structure and the actuator behaviour gives the operational point, i.e. the force and displacement for a given piezo stack actuator acting against an external spring. Maximum work is done when the stiffness of the actuator and external spring are identical.

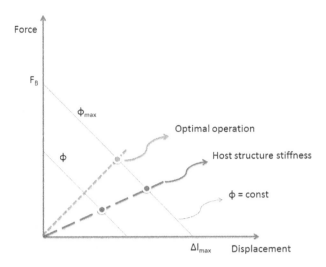

Figure 5.7. Operational behaviour of a perfect stack actuator

A numerical example of dynamic displacement of an unloaded piezoelectric stack actuator driven by a sinusoidal driving voltage is given in Fig. 5.8. The magnitude of the excitation signal at a frequency of 47.8 Hz is shown in Fig. 5.8(a). The resulting displacement of the piezo actuator is plotted in Fig. 5.8(b).

5.4 Smart Window Design

One of the main objectives that must be considered in an experimentally-oriented research is the easiness in the validation of the conceptual design. The overall simplicity and modularity of a research demonstrator can be as important as the performance it achieves.

It is commonly known that commercial aircraft windows can be rectangular with rounded corners, circular, elliptical or oval in shape. They cannot be square because this would cause high stress levels in the corners resulting in cracks and structural failures.

In this application, such geometrical characteristics are much simplified. The window is modelled and manufactured rectangular and flat for ease of fabrication and testing. Obviously, the resulting design is a strongly simplified model of an actual aircraft window. Nevertheless, this assumption is consistent with the literature available for double- and triple-wall partitions, largely addressing theoretical and experimental aspects in the control

5.4. SMART WINDOW DESIGN

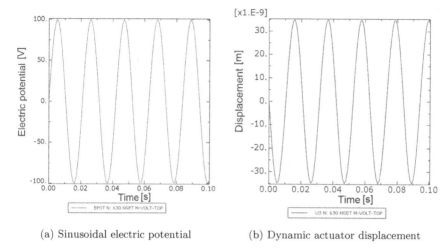

(a) Sinusoidal electric potential   (b) Dynamic actuator displacement

Figure 5.8. Dynamic response of a piezoelectric stack actuator driven by a sinusoidal voltage

of sound transmission through homogeneous and flat-panel systems. The proposed smart window is thus intended to be sufficiently representative of a commercial aircraft window, with the purpose of being a reference experimental test-bed aiming at comparing different active control techniques applied to a transparent multi-panel partition. Although the achieved results can be far from those obtained for a more realistic structure, they can be validated against analytical models of sound transmission through acoustically coupled flat multi-wall panels. Moreover, different strategies for active control of transparent panel partitions can be experimentally investigated under the same simulation framework.

A failsafe design of a dual pane glass window for commercial aircraft use is presented in [172]. A dual pane glass configuration has an advantage over plexiglass windows in that more stringent acoustic requirements can be met. Nevertheless, the authors conclude that in the event of structural failure, a further panel made of plexiglass or a laminate of plexiglass and glass would be strongly recommended for safety reasons.

In light of these considerations, a triple wall structural architecture is here considered. Besides the possibility of withstanding specific loads in case of failures, this solution enables breaking the vibro-acoustic chain through the partitions, giving significant benefits in the sound transmission loss behaviour. As demonstrated in Section 3.4.2, a change in density of the media travelling through the partitions separated by air spaces, results in a greater transmission loss than would have been provided by doubling the mass or thickness of a single partition in the same amount of space.

The design guidelines are taken from the theoretical study presented in [54]. The target noise level is 75 dBA as a maximum. As the fundamental blade passage frequency of the propeller aircraft lies around 124 Hz and the A-weighting reduces the sound pressure level at that frequency of about 16 dB, then a noise reduction of at least 49 dB would be required if the exterior overall sound pressure level impinging the window is 140 dB. Converting noise reduction to sound transmission loss requires the knowledge of the absorption coefficient at the frequencies of interest. The relationship between noise reduction σ and transmission coefficient τ is given by Eq. (5.8)

$$\sigma = 10 \log \left[1 + \frac{\overline{\alpha}}{\tau}\right]. \qquad (5.8)$$

For an average absorption coefficient $\overline{\alpha}$ of 20% at the blade passage frequency, the difference between noise reduction and transmission loss is approximatively 7 dB. This means that a transmission loss of 56 dB is required to obtain a noise reduction of 49 dB at 124 Hz for an exterior noise level of 140 dB. In this application, such a requirement is met by combining passive sound insulation with active noise control technology.

In contrast to most earlier works on panel control, mainly involving interior trim panels, the active structural acoustic control strategy is here applied to a multi-panel window by using a digital control system and integrated smart multi-functional piezoelectric materials. This results in a novel concept for interior noise control in aircraft cabin. The enhanced functionality of the proposed system will be tested through real-time experiments. The vibroacoustic performance of the smart window concept will be also evaluated in terms of comfort at passenger level.

5.4.1 *Scope*

A scheme of the proposed control concept is shown in Fig. 5.9. The time-advanced information on the airborne sound transmission through the window can be derived from the blade passage frequency (BPF) of the propeller, as demonstrated in similar applications [46]. A "comfort bubble" with reduced local NVH is realized by combining actively controlled aircraft windows with enhanced transmission loss properties. This target is, however,

5.4. SMART WINDOW DESIGN

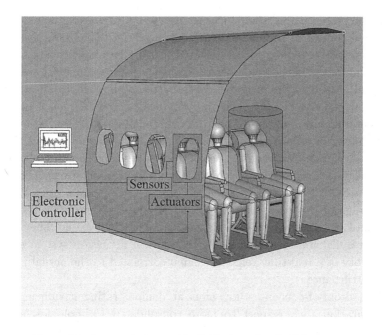

Figure 5.9. Smart window and individual comfort bubble concept

not totally new. The human-centred design of aircraft cabin environment has been addressed in many European research projects, such as SEAT [173]. The growing interest of aviation industry in this research field has led to innovative concepts for interior comfort such as the interactive window system developed for the new Boeing 787 Dreamliner aircraft [174]. Similarly, Airbus has announced research plans for a passenger airplane with a see-through light-weight intelligent fuselage, made of hi-tech ceramic skin, making the fuselage transparent when electrified [175]. Both examples reflect the new aircraft manufacturers interest in more customer oriented applications.

To date, only few control concepts have been realized for transparent multi-wall aeronautic structures and partially addressing acoustic problems. Currently, the most advanced solution involving aircraft windows for enhanced passenger comfort is available from Alteos® and consists of an innovative window system, larger in area than conventional windows, capable of changing light transmittance [174]. In other fields, Feonic® developed glass panel speakers actuated by smart materials for use in multiple applications [176].

In the present application, focus is given to multi-pane windows for interior noise control in commercial turboprop aircraft. The attractiveness of the proposed technology is investigated in a limited application scenario where the control of the fundamental frequency requires the most actuation. The goal is to develop an active control strategy applied to an advanced aircraft window in order to effectively reduce the low-frequency propeller noise entering aircraft cabin. The work concentrates on the frequency range below 200 Hz in order to cover the blade pass frequency at around 124 Hz. This poses a great challenge for the control strategy and the controller hardware due to the increased wavelength and amplitude oscillations of the noise disturbance.

Two major challenges need to be addressed in order to make the development of an adaptive noise blocking aircraft windows feasible. They are the need for a distributed and optically transparent actuation system and the availability of a real-time reference signal correlated to the actual primary noise disturbance.

This proof-of-concept study aims at demonstrating a sufficient level of maturation with respect to more consolidated technologies, such as loudspeaker-based ANC strategies. Following the NASA's technology readiness scale, a TRL 3-4 is expected as a minimum target. This implies a proof-of-concept validation in a laboratory environment. As always in the case of active noise control applications, further refinements and realistic testing are needed prior to their being incorporated into aircraft fuselage panels for the final demonstration in a relevant environment.

5.4.2 *Numerical modelling*

The triple-panel active window test-bed consists of three parallel and transparent window panes with an area of 32×20 cm, an aluminium frame and two enclosed cavities. Two of the three panes are made of 0.94 cm thick glass and are separated by an airspace d_1 of 1.66 cm. The third pane, which is made of acrylic, is 0.64 cm thick and it is separated from the closest glass pane by a 3.62 cm airspace d_2. The material properties are listed in Table 5.2. The triple-partition is asymmetric due to the difference in thickness. The panels are supported by strips of elastomer, as illustrated in Fig. 5.10. They can freely rotate. A silicon rubber is used to secure the window in the aluminium frame.

Figure 5.10 shows a section of the assembled triple wall structure, including the partitions and the strips of elastomer. A sketch of the window frame

5.4. SMART WINDOW DESIGN

Table 5.2 Material properties of the triple panel window

Material	Young's modulus (GPa)	Poisson's ratio	Density (kg/m^3)
Plexiglass	3.1	0.40	1190
Glass	62	0.24	2480
Elastomer	0.025	0.49	1300
Aluminium	72.4	0.33	2780

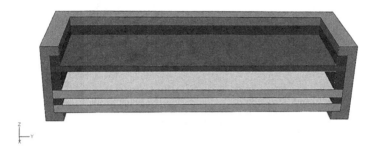

Figure 5.10. Cross-section of the triple panel system

and partitions is shown in Fig. 5.11. While mounted, the acrylic panel faces the inside of the fuselage cabin.

The FE model of the triple-wall structure is illustrated in Fig. 5.12. The two glass and acrylic panes, the elastomer and the aluminium frame are individually modelled with constant material properties. They are listed in Table 5.2. No variation with frequency in the damping properties of the elastomer is considered.

The aluminium frame, the glass and plexiglass partitions and the elastomer gasket are modelled with linear quadrilateral elements of 4-node general-purpose S4 shells, as shown in Fig. 5.13. The aluminium box is clamped into an infinite rigid baffle, as illustrated in Fig. 5.12(b). The surfaces of fluid not in contact with the structural partitions are assumed to be perfectly reflecting.

Before analyzing the sound transmission loss behaviour of whole system, the modal characteristics of each partition were studied individually. The natural frequencies of the panels were calculated analytically by assuming simply supported boundary conditions. Such frequencies are tabulated in Table 5.3. Then, the modal analysis of the triple wall model was carried out

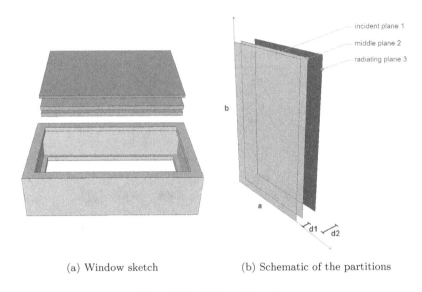

(a) Window sketch (b) Schematic of the partitions

Figure 5.11. Triple-wall window

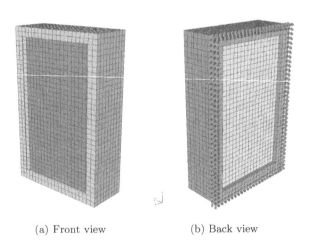

(a) Front view (b) Back view

Figure 5.12. FE model of the triple-wall structure

in ABAQUS®. Natural frequencies and mode shapes were computed in the frequency range up to 800 Hz.

As shown in Table 5.3, the window prototype and the simply supported plexiglass panel show comparable resonance frequencies. Furthermore, the numerical results are in agreement with the analytical predictions. This

5.4. SMART WINDOW DESIGN

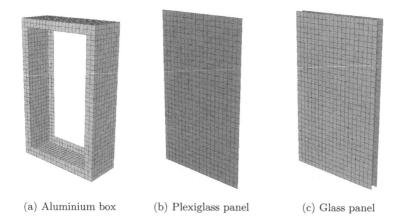

(a) Aluminium box   (b) Plexiglass panel   (c) Glass panel

Figure 5.13. FE model of the window components

Table 5.3 Calculated window resonance frequencies

	Analytical				Numerical	
Resonance system	Material	Thickness (cm)	Mode	Simply supp. (Hz)	FE model (Hz)	Structure-acoustic FE model (Hz)
Single layer	Plexiglass	0.64	(1,1)	177.58	175.77	
			(1,2)	327.23	324.65	
			(2,1)	560.69	555.15	
			(1,3)	576.65	570.75	
	Glass	0.94	(1,1)	762.85	754.27	
			(1,2)	1405.7	1389.3	
Window prototype			(1,1)		170.48	194.17
			(1,2)		304.76	297.52
			(2,1)		475.37	470.36
			(1,3)		528.19	526.09
			(2,2)		612.17	606.80
			(2,3)		804.58	796.52
Cavity			(0,1,0)			538.06

suggests that the boundary conditions with the panes supported by the elastomer, as shown in Fig. 5.10, are approximated closely by simply supported conditions. This is also demonstrated by the vibro-acoustic coupled analysis of the simplified triple wall structure described in Section 3.4.2. The coupled modal resonances computed for the simply supported triple-wall partition,

164   CHAPTER 5.   NOISE-REDUCING SMART WINDOWS

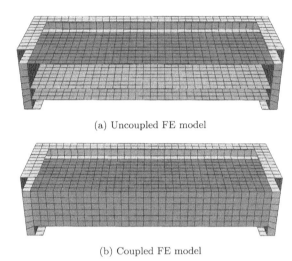

(a) Uncoupled FE model

(b) Coupled FE model

Figure 5.14. Air gap modelling

listed in Table 3.5, match those predicted by the FE model of the triple-glazed window, shown in Table 5.3.

Being a triple-leaf partition, particular attention has been paid to the simulation of the air gap between the plexiglass and the upper glass layer and that between the upper and lower glass layers. A coupled FEM-FEM analysis was performed in ABAQUS®. The air between the plexiglass and the glass layer was modelled with 2560 linear hexahedral elements of type AC3D8, while 1280 AC3D8 elements were used to fill in the volume of the interior air between the two glass layers, Fig. 5.14. To simplify the problem, the effect of the air outside the window box was neglected. The air was modelled with a density of $1.225 \,\text{kg/m}^3$ and a bulk modulus of 0.142 MPa. The surface-based contact approach was used to couple the structural and acoustic meshes. Such coupling surfaces were defined for the frame, the opposite layers and the enclosed air gap. The TIE option was used to couple the structure with the inside air by assuming the acoustic medium tied to the structure. Boundary conditions of $v_n = 0$ were prescribed on all surfaces of the fluid not in contact with the structure, with v_n being the normal velocity on the fluid surface. As a result, the *in vacuo* structural modes and the acoustic pressure modes of the rigid-walled cavities were complemented by a new kind of dynamic behaviour, in which the system's vibration involved both the shell and the fluid regions. These modes are referred to as acoustic-structure resonances.

5.4. SMART WINDOW DESIGN

A comparison between the natural frequencies resulting from the uncoupled and coupled FE analysis is given in Table 5.3. The first six mode shapes of the triple-wall structure are reported in Fig. 5.15. Notice how the structure-acoustic coupling affects the dynamic behaviour of the vibro-acoustic system. Due to the coupling effects, the eigenfrequencies shift mode by mode. The first natural frequency of the coupled system is much higher than the *in vacuo* structural mode. This is mainly due to the acoustic chamber, represented by the fluid between the panes, which "spring-loads" the two plates.

Figure 5.16 shows the sound pressure distribution of the acoustic medium within the cavity. Due to the boundary conditions, the pressure distribution in the air gap between the plexiglass and the upper glass layer (left side of the picture) matches the first bending mode (1,1) of the plexiglass plate (right side of the picture), Fig. 5.16(a). The same behaviour is observed also for the second bending mode (1,2), as shown in Fig. 5.16b. The frame components of the first mode shape are shown in Fig. 5.17. Finally, the cavity resonance occurring in the air volume comprised between the plexiglass and the upper glass layer is shown in Fig. 5.18.

5.4.3 Structural control actuators

In order to beneficially control the inherent structural characteristics of the window, an adaptive structural system is built from the passive configuration by integrating piezoelectric actuators. Since the main interest lies on controlling the mass–air–mass resonance of the triple-wall structure, piezoelectric actuators with enough control authority over the low frequency range are required. Moreover, if the view through the transparency shall be preserved, such actuators must not interfere with the primal function of the controlled structure, namely to look through.

The actuation mechanism of the smart window consists of ten piezoelectric stack actuators placed at the borders of the radiating panel, as illustrated in Fig. 5.19. This configuration allows for decoupling the incident and the radiating panel vibrations as well as controlling the flexural modes more effectively by controlling the driving voltage applied to each actuator individually [177]. These COTS actuators are distributed along the window frame in order to provide in-plane forces on the radiating panel of the window by taking advantage of the d_{33} effect. Moreover, from a more practical point of view, they can be easily hidden in the frame of the window. Figure 5.20 shows a schematic of the actuation mechanism for a single actuator exciting the radiating panel. In this configuration, each piezo actuator generates an

166   CHAPTER 5.   NOISE-REDUCING SMART WINDOWS

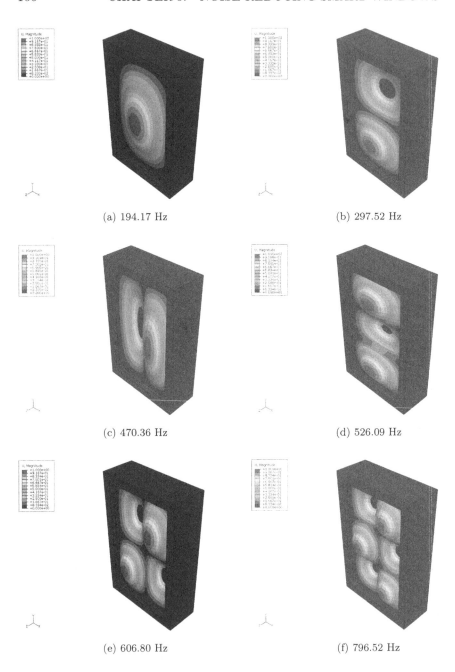

Figure 5.15. Numerical modes of the radiating panel of the triple-wall structure

5.4. SMART WINDOW DESIGN

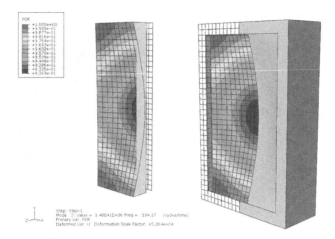

(a) Section of the pressure distribution in the air gap between the plexiglass and glass panels – 1st mode

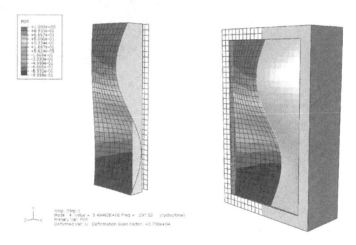

(b) Section of the pressure distribution in the air gap between the plexiglass and glass panels – 2nd mode

Figure 5.16. Structure-acoustic modes

off-midplane actuation force on the radiating panel inducing local bending moments aiming at controlling the out-of-plane vibro-acoustic modes. These dynamic forces are transferred to the plexiglass panel through metal rods, which keep the piczoceramic actuators in their compressive state when

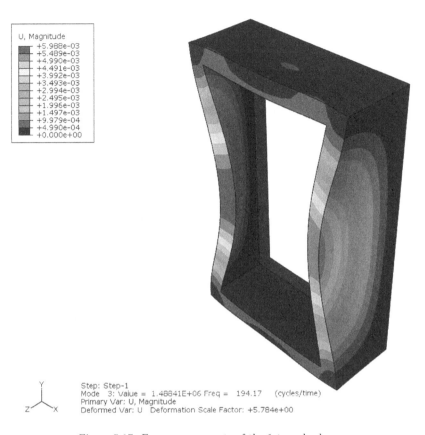

Figure 5.17. Frame components of the 1st mode shape

no voltage is present. When a positive electric potential is applied to the electrodes, the piezoceramic material geometrically expands, causing the rod to move forward in order to accomodate for the increased length of the piezo actuator. By applying the actuators with a certain eccetricity with respect to the middle plane of the radiating panel, the motion of the d_{33} piezo actuator results in a concentrated bending moment, forcing the structure in the out-of-plane motion. Unlike the laminar piezo which transforms the d_{31} effect in the plane of the actuator to an out-of-plane motion by gluing the actuator to the structure, this configuration applies concentrated forces at the boundaries to realize a bending action.

A sketch of the segmented control architecture is shown in Fig. 5.21. Although the area in direct proximity of the boundaries is not very effective

5.4. SMART WINDOW DESIGN

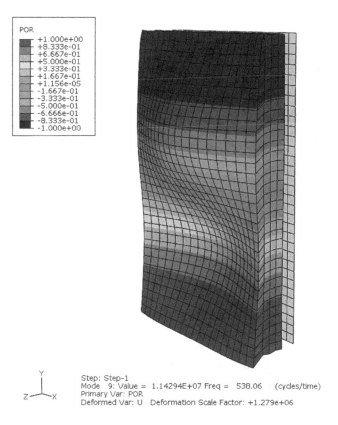

Figure 5.18. Cavity resonance in the air gap between the plexiglass and glass panel

in exciting flexural modes, which are responsible for sound radiation, this adaptive interface is particularly suited for transparent partitions also preventing the transmission of structure-borne sound. Due to the passive barrier, represented by the two glass plates located at the incident side of the triple partition, a lower voltage actuation is required to control of the out-of-plane vibration of the radiating panel. Furthermore, only the lower-order radiation modes need to be controlled to achieve a substantial reduction in the radiated power due to their higher radiation efficiency. By driving the actuators in phase, the piston-like coupled mode is then controlled efficiently. If the actuators are individually controlled, the control of sound transmission can be extended over a wider frequency band.

The high voltage PI® P-010.10P PICA Stack Power actuators are selected for this application, Fig. 5.22. These actuators are specifically

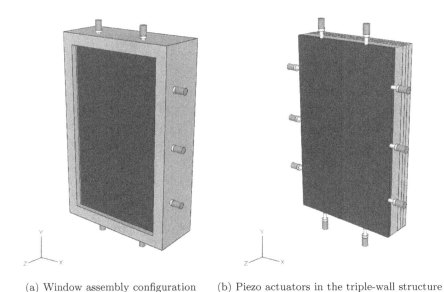

(a) Window assembly configuration (b) Piezo actuators in the triple-wall structure

Figure 5.19. Structural model of the smart window prototype

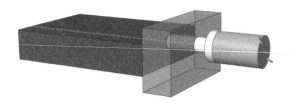

Figure 5.20. Sketch of a single actuation system

designed for active noise and vibration control applications since they reach the nominal response in approximatively 1/3 of the period of the resonant frequency. The technical datasheet is reported in Table 5.4 [178]. The actuators consist of thin layers of electroactive ceramic material which are electrically connected in parallel. Their combination of high displacement and low electrical capacitance provides a reliable dynamic behaviour with reduced driving power requirements. The thickness of the individual layers determines the maximum operating voltage of the actuator. The maximum electrical field that is recommended for effective applications is in the order of 1 kV peak to peak. The output of the controller is thus connected to the high voltage PI® E-471 High Power amplifier [179].

5.4. SMART WINDOW DESIGN

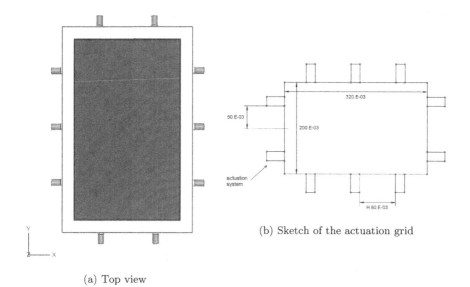

(a) Top view

(b) Sketch of the actuation grid

Figure 5.21. Configuration for piezo stack actuators

Figure 5.22. P-010.xxP PICA Stack Power actuators [178]

Table 5.4 Technical data of the P-010.10P PICA Stack Power actuators [178]

PI® PICA Stack Power actuators	
Model	P-010.10P
Displacement [μm −10/ + 20%]	15
Diamater [mm]	10
Length [mm]	18
Blocking force [N]	1800
Stiffness [N/μm]	120
Capacitance [nF ± 20%]	46
Operating temperature range [°C]	−20 to 150
Operating Voltage range [V]	0–1000
Resonant frequency [kHz]	64

An index of the piezo actuator capability to work at a specific frequency is represented by the thermal dissipation factor. The thermal active power generated by a piezo actuator during harmonic excitation is given by

$$W = \frac{\pi}{4} \tan(\delta) C V^2 f \tag{5.9}$$

where $\tan(\delta)$ is the dielectric factor (≈ 0.1), C is the actuator capacitance, V is the peak-to-peak voltage and f is the operating frequency. The dissipation of the generated power is dependent on the actuator volume, with an acceptable limit of 1 Watt per cubic centimeter.

The high power actuator performance degrades for increasing frequencies. The actuator can be fully driven up to 200 Hz while maintaining a proper functioning. At that frequency the dissipation factor at 1000 V is equal to 0.51 W/cm^3. The corresponding unconstrained displacement is 15 μm while the blocked force is the nominal value. At 500 Hz the dissipation factor reaches 0.923 W/cm^3 for a driving voltage of 850 V. This results in a slight reduction in the maximum force (blocked force) as well as in the maximum unconstrained displacement.

Preliminary experimental studies on the use of piezo stack actuators are described in Appendix C. The effectiveness of the actual control mechanism in exciting the structural modes of the radiating panel has been experimentally evaluated. The results of these tests are discussed in Section 5.5.2.

5.5 Smart Window Test-Bed

The fully assembled and functional smart aircraft-type window test-bed is shown in Fig. 5.23. It consists of an aluminium frame containing three flat

5.5. SMART WINDOW TEST-BED

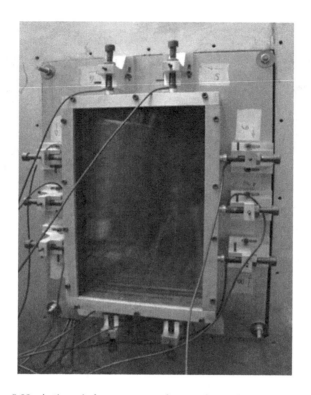

Figure 5.23. Active window prototype for an advanced turboprop aircraft

transparent panels with area 20 × 32 cm. The overall dimensions are 22.2 × 34.2 × 8.72 cm. The total weight sums up to 3.3 kg without taking into account cables for power supply of sensors and actuators. In Fig. 5.24a, the dimensions of the test-bed are qualitatively compared with those of an actual aircraft-type window of the ATR 72 aircraft. However, the latter, shown on the right, does not include the inner acrylic panel usually comprised in the aircraft interior lining panels, as shown in Fig. 5.24b.

The piezo stack actuators are integrated into the window prototype via the active attachment elements. The planar actuation system is constrained to the window frame by a dedicated system of clamping, shown in Fig. 5.25. The mounting elements are manufactured in thick aluminium. They provide a support that holds the actuator end when the device tries to expand. Screws up to a size of M10 provide a preload for the actuators. This preload allows to prevent the actuators to operate in tension, a condition that could damage the piezoceramic material.

174 CHAPTER 5. NOISE-REDUCING SMART WINDOWS

(a) Smart window test-bed

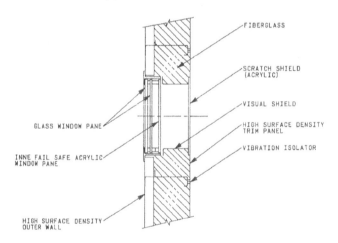

(b) Sketch of window installation in a double-wall fuselage [54]

Figure 5.24. Smart window demonstrator

5.5.1 *Experimental modal analysis*

The numerical modal model of the structure was experimentally validated by performing an impact modal test, Fig. 5.26. The advantage in using this experimental technique is that it allows a quick estimation of the modal parameters by processing the FRF resulting from a time-limited excitation

5.5. SMART WINDOW TEST-BED

Figure 5.25. Clamping system of the piezo actuators

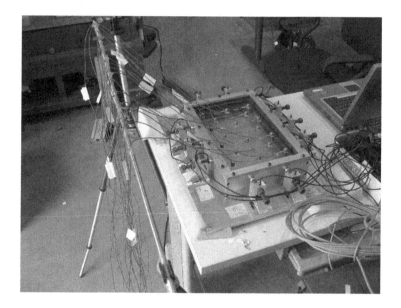

Figure 5.26. Experimental modal analysis of the window prototype

of the structural modes in the frequency range of interest. In Section 6.4, the analysis of the dynamic behaviour in operational conditions is extended to a stepped sine excitation provided by driving the piezoelectric actuators.

A schematic representation of the impact test is given in Fig. 5.27. The open-loop (i.e. no control) modal behaviour of the prototype was characterized in the frequency range up to 600 Hz. An instrumented hammer was equipped with a load cell aiming at measuring the force induced to the system. Eighteen accelerometers were distributed over the structure with a regular mesh spacing in the central part of the plexiglass panel, Fig. 5.28.

176   CHAPTER 5. NOISE-REDUCING SMART WINDOWS

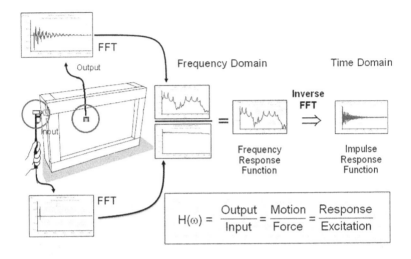

Figure 5.27. FRF measurements through impact testing

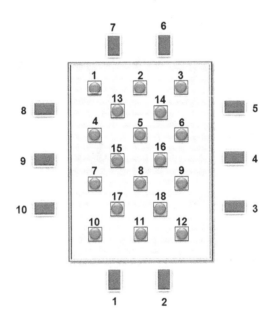

Figure 5.28. Experimental grid of measurement and actuation

5.5. SMART WINDOW TEST-BED

This network of acceleration sensors were aimed at measuring the dominant low-frequency radiating mode shapes. The signals measured by the load cell as well as the acceleration levels were conditioned and digitalized through the LMS® SCADAS III data acquisition system and finally processed by the LMS® Test Lab software. The data acquisition was carried out by using a sample frequency of 16 kHz.

The experimental set-up is illustrated in Fig. 5.29. Light weight piezoelectric accelerometers with integral electronics were used. The Endevco 7259B-10 model (4.6 gm), shown in Fig. 5.29c, was used to effectively minimize mass-loading effects due to the sensors. The structure was tested clamped on a reference panel resting on a block of foam rubber simulating free-free boundary conditions.

The FRFs resulting from impact loads were measured first. Next, the modal parameters were extracted from the stored FRFs by applying the LMS® Polymax method. Figure 5.30 shows two experimental FRFs measured by two sensors located nearby the centre of the plexiglass plate. As numerically simulated, no prestress on the piezo actuators was imposed through the screws. The first two experimental mode shapes are shown in Fig. 5.31. The experimental modal frequencies as well as the numerical-experimental correlation results are tabulated in Table 5.5. The numerical results match very well the experimental ones.

The influence of the mechanical prestress onto the structure was finally examined. To this aim, a uniform torque was applied to the screws constraining the piezo actuators through the borders. As shown in Fig. 5.32, a slight reduction in the resonance frequencies of the structure was measured depending on the amplitude of the resulting mechanical load. The resonance frequencies decreased for increasing compression preloads. This was obtained for two different levels of torque (1 Nm, 2 Nm) tightening the screws.

5.5.2 Control actuators set-up

Reliable numerical predictions and extensive experimental campaigns, aiming at validating the main technological aspects related to sensing and control capabilities, are crucial factors for an efficient control of piezo-based systems. The optimization of the number and size of multiple actuators with regard to a given host structure can be very effective in reducing radiated sound power. If the structure is supposed to be actively controlled, actuators and sensors must be able to excite and sense the modal oscillations, respectively.

178 CHAPTER 5. NOISE-REDUCING SMART WINDOWS

(a) Experimental set-up

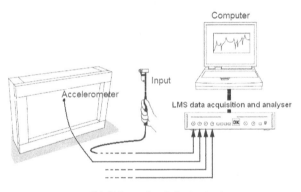

(b) Schematic of the test set-up

(c) Endevco 7259B10 transducer

Figure 5.29. Experimental set-up

If a force actuator is unable to excite a mode, then that mode is said to be not controllable.

Experiments were performed in order to investigate structural modal excitability through single actuators as well as grouped piezo stacks. Focus was given to the out-of-plane vibrations as well as to the acoustic radiation properties of the triple wall structure when excited by piezo stack actuators driven by a broadband driving signal (sine sweep, 100 V of amplitude).

5.5. SMART WINDOW TEST-BED

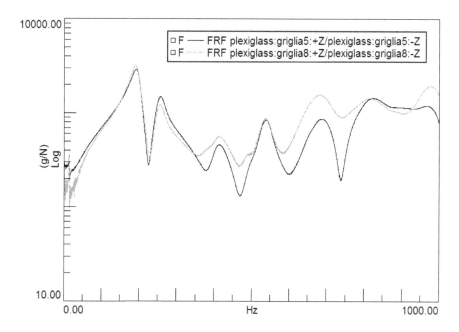

Figure 5.30. Experimental FRFs for modal analysis

The experimental evaluation of the actuation capability was carried out by mounting the window prototype on a reference test suite. More details on the experimental set-up are given in Chapter 6. Accelerations normal to the plate surface were measured on a test grid, shown in Fig. 5.28. Frequency response functions (FRF) were computed from the ratio between the Fourier transform of the acceleration signals measured on the acquisition points and the reference signal driving the piezo actuators. Figure 5.33 shows the averaged FRF obtained by driving all the actuators concurrently and with the same phase and amplitude. The actuator positions are schematically shown in Fig. 5.33b. The mean acceleration level is defined as the sum of the squared accelerations measured by each sensor. In case of in-phase excitation, the structural modes included in the frequency range up to 500 Hz were sufficiently excited. For this configuration, only the [odd, odd] modes were effectively controlled by the piezo actuators acting on the borders of the structure. This implies that in circumstances where the sound waves of an external disturbance incident on the panel partition are not perpendicular plane waves, the control performance will certainly degrade since some of the [odd, even], [even, odd] and [even, even] modes will be significantly excited. Nevertheless, with an appropriate excitation of all actuator

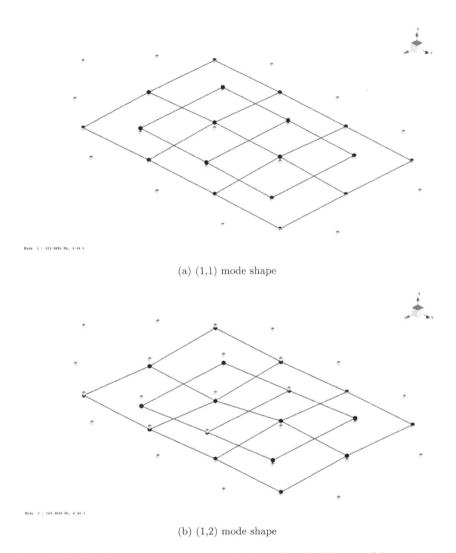

(a) (1,1) mode shape

(b) (1,2) mode shape

Figure 5.31. The first two experimental mode shapes identified by a modal separation technique

units, enhanced performance can be reached depending on the amplitude and phase driving each piezo actuator. In a multi-output active control system, the driving voltage applied to each actuator can be optimized in order to effectively suppress single eigenmodes with higher radiation efficiency or to excite additional modes in such a way that the overall acoustic radiation behaviour becomes less efficient. The (1,1) mode and (1,2) mode of the

5.5. SMART WINDOW TEST-BED

Table 5.5 Numerical-experimental comparison

Mode	Uncoupled FEM (Hz)	Coupled FEM (Hz)	Experiments (Hz)	Error (%)
(1,1)	170.48	194.17	197.00	1.43
(1,2)	304.76	297.52	285.00	4.39
(2,1)	475.37	470.36	455.00	3.37
(0,1,0)		538.06	536.00	0.4

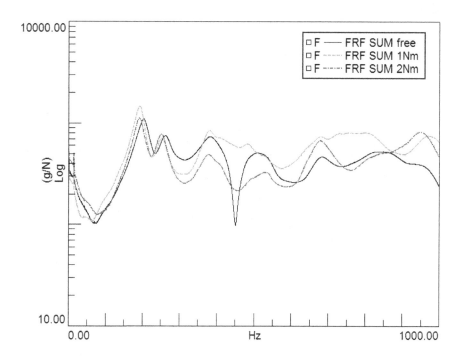

Figure 5.32. Influence of the compressive preload on the structural response

smart window prototype were effectively excited by driving the actuators 5 and 10 in-phase and with the same magnitude and by shifting of 180° the phase of actuators 2 and 7. The first pair is located either at the left or at the right side of the demonstrator. The second either at the top or at the bottom. This is experimentally demonstrated in Fig. 5.34. A detailed view of the actuator mounting is given in Fig. 5.35. On the other hand, the modes of the active window test-bed were poorly excited if the number of control actuators decreased. Figure 5.36 shows the FRF measured by driving only the actuator 2. Both the (1,1) mode and the (1,2) mode were not clearly

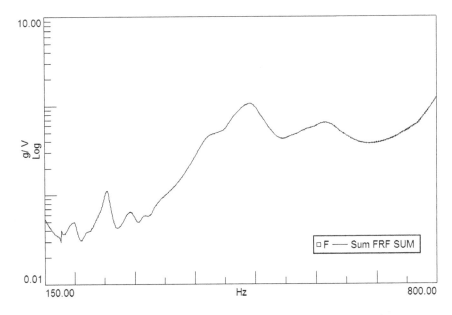

Figure 5.33. Fundamental FRF averaged over the measurement points

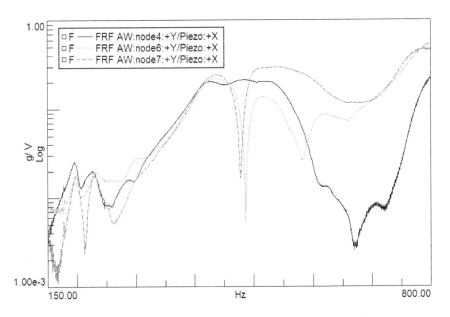

Figure 5.34. Four integrated actuators for mode (1,1) and mode (1,2) excitation

5.5. SMART WINDOW TEST-BED

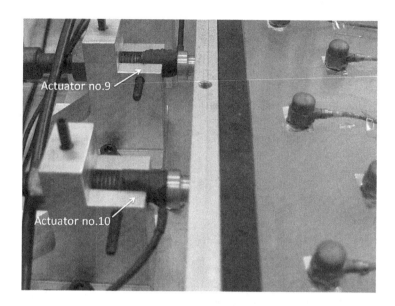

Figure 5.35. Detail of the actuator mounting system

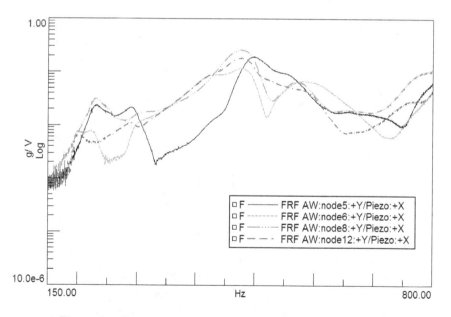

Figure 5.36. Experimental FRFs measured on points 5, 6, 8 and 12

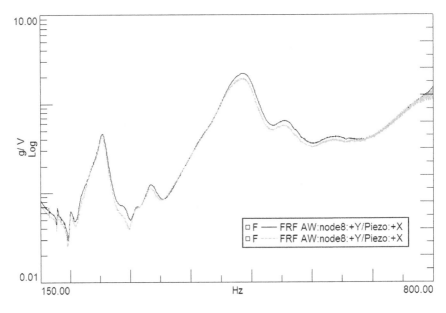

Figure 5.37. Excitation of mode (1,2) by driving two actuators

Figure 5.38. Experimental set-up in anechoic chamber

5.5. SMART WINDOW TEST-BED

detected and distinguished by the sensor network. The experiment was then repeated by adding the actuator 7. Figure 5.37 shows the resulting FRF measured on node 8. To excite the (1,2) mode, the actuators were driven in opposite phase. This experiment was performed twice for repeatibility reasons. The two curves are in good agreement.

Finally, the low-frequency acoustic relevance of the structural modes was investigated. In order to derive a preliminary evaluation of the critical modes which may impact on the performance of the active noise control system, the free field sound radiation was determined experimentally, Fig. 5.38. An experimental campaign was carried out inside an anechoic chamber by driving all the actuator units simultaneously. The window prototype was tested as an acoustic speaker. An input voltage of 300 V (peak to peak) was supplied to the piezo actuators in the frequency range 100–600 Hz by means of a voltage amplifier. The output of a pressure transducer was sampled and processed with a spectrum analyzer. Figure 5.39 shows the frequency response function measured by a microphone placed at a distance of 1 m from the window prototype. Experimental results show that the low-frequency acoustic behaviour of the structure is dominated by the first structure-acoustic mode. This is demonstrated by the resonant peak located around 200 Hz which is expected

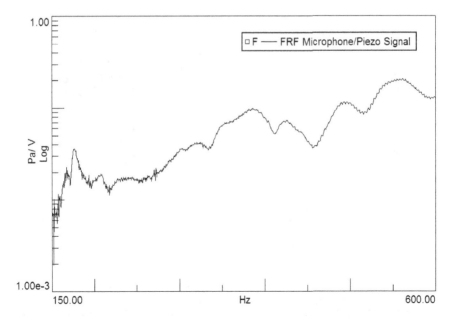

Figure 5.39. Low-frequency acoustic radiation of the window prototype

to dominate the sound transmission loss behaviour at low frequencies. This is in agreement with the numerical predictions. Contrary to the first vibration mode, the (1,2) mode behaves like a less efficient sound radiator due to the low net space-average volumetric velocity forcing the surrounding fluid to vibrate.

5.6 Transmission Loss Predictions

The sound transmission loss capabilities of the passive triple-wall window test-bed were predicted by applying the simulation tool detailed in Chapter 3. Contrary to the simply supported partitions discussed in Section 3.4.2, the panels of the triple-wall test-bed are physically connected by the window frame and the supporting system. These mechanical connections provide a further structural path promoting the sound transmission between the partitions. As shown in Fig. 5.40, the impact of the structure-borne sound path on the sound reduction index is particularly important at frequencies above 300 Hz. At those frequencies, the primary sound transmission path

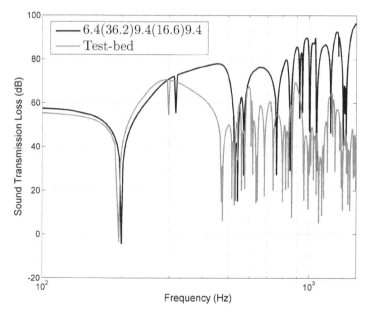

Figure 5.40. Transmission loss comparison between the triple-wall window test-bed and an equivalent simply supported triple-panel partition

is from pane to pane through the edge connections. On the other hand, a strong coupling between the structure and the enclosed fluid occurs below 200 Hz. The plexiglass panel radiates in a very efficient mode resulting in a local transmission loss dip at 194.17 Hz.

It should be noted that the sound transmission loss behaviour of the triple-panel structure meets the interior noise targets mentioned previously. Nevertheless, since the main discrepancies with respect to the simply supported triple-panel partition are due to the structure-borne sound path, better transmission loss performance can be achieved if the partitions are not mechanically connected or the vibration levels are actively controlled. Therefore, active noise control will be considered in the following chapters to further enhance the low-frequency transmission loss properties of the window test-bed.

5.7 Summary

The analysis of the state-of-the-art allows us to establish some interesting conclusions on the active control of multiple partitions. The use of vibration control sources offers two advantages over the use of more conventional acoustic sources: (i) fewer secondary sources are generally required for global control of the interior noise field and (ii) surface mounted actuators are less intrusive than loudspeaker arrangements. On the other hand, for cavity-dominated modes, the use of secondary loudspeaker and error sensors inside the air cavity between the partitions represents a more effective solution than controlling the vibration of one of the plates. However, the large number of microphones required to achieve global noise control may affect the practical system compactness.

In this chapter, some background information has been given on piezo-electricity. The basic equations governing the coupled electromechanical behaviour of such materials have been introduced. In the detailed application, the use of piezoelectric actuators for active structural acoustic control is restricted to piezoelectric stacks.

Subsequently, an actively controlled aircraft-type window test-bed, used as benchmark for theoretical comparisons throughout the contents of this book, has been presented. A detailed FE model of the coupled structural-acoustic system, including the fluid between the panels, has been developed as a basis for the control system design and performance evaluation. The numerical model has been successfully validated through an experimental modal test. The modal parameters have been extracted by using a phase separation technique. A comparison has been made between the numerical and experimental results by focusing on the coupled structural mode shapes of the window assembly at low frequencies.

Focus has been given to the structural modal excitability achieved by driving the piezoelectric actuators. A dedicated experimental campaign has been performed in order to investigate the actuation effectiveness. The experimental results have demonstrated that the low frequency modes have been sufficiently controlled with a reduced set of vibration sources. It must be noted that the "open-loop" system has been considered. This means that the active control system has not been included in this chapter. ASAC simulations will be presented in the next chapter. Active control of sound radiation involves the superposition of secondary control actuators introducing counteracting forces aiming at reducing the residual sound field.

Finally, the proposed simulation tool for predicting the sound transmission loss behaviour of plate-like multi-wall structures has been applied to the triple-wall window test-bed. A good matching between the sound reduction index of the window test-bed and that of a simply supported triple-panel partition has been achieved, although the former has shown characteristic resonant effects. Apart from the medium confined in the acoustic cavities, the primary transmission path has been influenced by the structure-borne sound path contribution.

Chapter 6

Active Noise Control Experiments

6.1 Introduction

Several research studies on the use of distributed piezoelectric actuators, mounted on either the incident or radiating panels of multi-panel partitions, have been conducted and reported in the literature in order to enhance the amount or transmission loss by taking advantage of both passive and active control methods [26, 27, 37]. Moreover, a large portion of the research on active structural acoustic control is concerned with the use of simplified models of aircraft fuselage shells simulating the transmission path from the exterior of the fuselage to the interior environment.

In this chapter, the effectiveness of actively-controlled smart panels in providing better noise attenuation than their passive counterparts is validated by experiments involving sensors, actuators and electronic devices. In this context, guidelines for active noise control experiments are also provided. The control system performance is evaluated under laboratory conditions. For a direct analysis, the smart window, including the actuation system, is installed between two acoustic chambers. A vibro-acoustic test facility is employed as sound transmission suite. In the experimental set-up, the acoustic excitation is realized by means of a loudspeaker. The local and global sound radiation is measured within an acoustic anechoic chamber, facing the radiating panel of the window. In such conditions, the sound flows directly out from the source, and both pressure and intensity levels drop with increasing distance from the source according to an inverse square law. As a result, the spatial radiation patterns can be estimated accurately without any reflection, as well as the amount of noise transmitted through or attenuated by the smart panels. The control capabilities of the smart window are then evaluated by comparing the sound transmission loss achieved with active control versus the original passive configuration. The measurement method as well as the control loop implementation and the experimental results are therefore detailed.

The active controller driving the smart window demonstrator is developed in MATLAB®/Simulink® block diagram environment. A C code is generated in Real-Time Workshop® and implemented in a dSPACE DSP control board for real-time experiments. For most experiments involving both SISO and MIMO control strategies, the sampling frequency is set to 9.5 kHz.

The active noise control results are described in terms of sound pressure levels at four pressure sensors located at different distances from the radiating side of the window panel. The sound power radiated in the anechoic chamber is also measured by sound intensimetry. The sum of the mean squared sound pressure, measured in control points, referred to as error microphones, is chosen as the index of the control objective. This criterion is proportional to the sound power radiated by the structure under a far-field assumption. A filtered-reference LMS approach is implemented for the adaptation of the control filters minimizing the quadratic controller objective function. A straightforward gradient descent method is employed to iteratively derive the optimal set of the controller coefficients. The piezo stacks actuators are therefore driven in order to adaptively minimize the effect of the incident noise field transmitted through the structure. Both the structural vibrations and the amount of noise radiated by the structure are real-time monitored.

In Section 6.2, the experimental set-up for the active control of sound transmission is presented. The low-frequency dynamic behaviour of the window prototype is evaluated in operating conditions. In order to analyze the vibro-acoustic coupling within the experimental facility, a coupled structure-acoustic FE model is presented in Section 6.4. The developed control strategy is described in Section 6.5, as well as its implementation on a DSP control board. The experimental results achieved for the single- and multi-channel controllers are also provided. Details on the sound intensity measurements of the radiating panel surface with and without active control are given in Section 6.6. The total sound power radiated by the structure is also determined. Finally, the human perception of noise and

6.2 Experimental Set-Up

Experiments were conducted with a laboratory set-up in order to demonstrate the potential of the proposed active structural acoustic concept. The experimental set-up was built at the Smart Structures and Vibro-Acoustic Laboratory of the Italian Aerospace Research Centre (CIRA). It consisted of two rooms, the sending room where the noise source was placed and the receiving room where the transmitted noise was measured. The receiving room was a semianechoic chamber with dimensions 6.5 m × 5.3 m × 4.2 m (volume 145 m^3) and a cut-off frequency of 70 Hz. The walls inside the chamber were lined with fiberglass wedges absorbing 99% of all acoustical incident energy within the frequency range. The background noise due to external sources was below the minimum audible field. The chamber was equipped with calibrated sound pressure microphones and a full instrumentation suite. The sending room was a rigidly walled steel box, shown in Fig. 6.1, having dimensions 1.2 m in height, 1.06 m in width and 0.79 m in length. It was made of 30 mm thick steel panels, interfacing each other by insulating rubber. The structure was constructed in such a way that it had

Figure 6.1. Vibro-acoustic facility, sending room

192   CHAPTER 6. ACTIVE NOISE CONTROL EXPERIMENTS

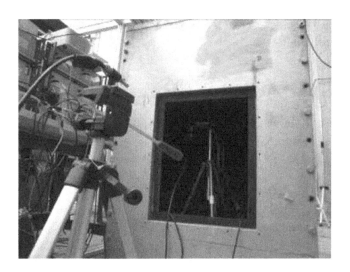

Figure 6.2. Transmission test window in the sending room

removable parts and its interior was accessible for changes in the acoustic treatments, as shown in Fig. 6.2. Moreover, the acoustic excitation source could be suitably positioned in its interior. It was supported by eight rubber mounts ensuring a cut-off frequency well under 10 Hz. The overall weight was about 2 tons.

Such a vibro-acoustic facility was designed, manufactured and tested in the framework of the BRITE-EURAM GASS research project for active noise control of aircraft interior trim panels. It was designed so that the insertion loss of the side walls of the box was at least 10 dB higher than that of the panel under study in the frequency range 0–500 Hz. The efficiency of the test box was previously verified by a series of sound intensity measurements. For the present study, the test facility was reassembled and slightly modified to accomodate the smart window panel in the wall opening between the two rooms.

Figure 6.3 shows the smart window prototype installed in the side opening of the vibro-acoustic facility. The window panel was mounted in the open lateral side of the box via four attachment points. A schematic view of the experimental set-up is shown in Fig. 6.4. The noise was generated in the sending room by a dodecahedron speaker, transmitted through the triple-wall structure and measured in the receiving room. The system is described as follows: the acoustic plane waves excite the incident panel of the triple-panel system. The induced plate motion excites the acoustic cavities, thus

6.2. EXPERIMENTAL SET-UP

Figure 6.3. Active window mounted in the side opening of the sending room

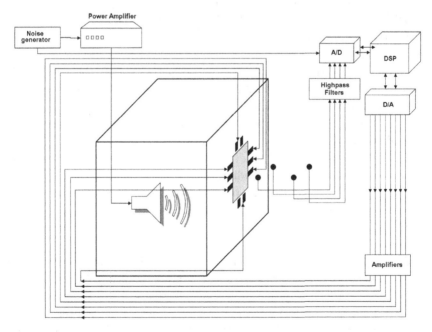

Figure 6.4. Schematic presentation of the experimental set-up with the smart window prototype mounted on the testing facility

194   CHAPTER 6. ACTIVE NOISE CONTROL EXPERIMENTS

Figure 6.5. View of the experimental set-up from the receiving room

inducing motion to the radiating panel. The radiating panel then emits sound power into an acoustic free field. In Fig. 6.5, the whole assembly installed into the anechoic chamber is shown. Figure 6.6(a) shows a view of the smart window from the anechoic chamber. The sound pressure levels were measured next to the window as well as at different distances. A seat was positioned in the receiving room in order to simulate the passenger's position. In Fig. 6.6(b), a view from the vibro-acoustic facility is illustrated. The incident panel corresponds to the aircraft exterior. The omnidirectional speaker, shown outside the vibro-acoustic facility in Fig. 6.5, is located in front of the incident panel.

In the laboratory set-up, the incident panel was associated with the outer aircraft fuselage and the radiating panel with the inner fuselage. As a result, the free space next to the radiating panel approximates the passenger's cabin environment, while the speaker simulates the external acoustic disturbance. The incident acoustic field was the noise field generated by the propeller of a turboprop engine. The radiated acoustic field was representative of the aircraft interior. It was assumed highly damped in order to experimentally characterize the acoustic power transmitted through the triple-wall partition by performing sound intensity measurements in the acoustical free field of the receiving room. As discussed in Chapter 5, active control was provided by piezoelectric actuators mounted on the radiating panel of the triple-wall structure.

In addition to the speaker, an array of 2×2 microphones was positioned inside the vibro-acoustic facility in order to monitor the generated sound

6.3. INSTRUMENTATION

(a) Monitoring microphones in the receiving-room, passenger position

(b) Monitoring microphones in the sending room

Figure 6.6. (a) View from the anechoic chamber and (b) view from the interior facility

field. Moreover, the facility was instrumented with four accelerometers, as shown in Fig. 6.13(a). The window prototype was also instrumented with 18 accelerometers mounted on the radiating plate, as shown in Fig. 6.6(a). A sketch of the experimental grid is depicted in Fig. 5.28. Four microphones were placed outside the vibro-acoustic facility in order to map the sound pressure radiation. They provided the input signals to the active noise control system. The exact position of the microphones used for the experiments is given in Table 6.2. The instrumentation used for the experimental set-up is discussed in Section 6.3. The experimental results of the single- and multi-channel real-time controller are detailed in Section 6.5.

6.3 Instrumentation

The experimental results presented in this book were obtained by using the following equipment:

- *Microphones* (PCB 377B02). The acoustic pressure in discrete field points was measured by half-inch prepolarized free-field condenser microphones

(50 mV/Pa), 3.15 Hz–20 kHz with ICP preamplifier (426E01) and TEDS. They were used as error and monitoring sensors for the active control system. During all the tests the microphone signals were conditioned by a PCB 16 channels ICP conditioner.
- *Accelerometers* (Endevco 7259B-10). Small lightweight accelerometers (4.6 gm, 10 mV/g) were mounted on the radiating panel in order to measure accelerations normal to the plate surface. The acceleration signals were conditioned by the data acquisition system.
- *Sound intensity probe* (B&K 3595). A two-microphone sound intensity probe with half-inch microphones (type 4197) was used to measure the sound power transmitted through the window prototype (more details on the technique are given in Section 6.6).

The primary noise field was generated by a dodecahedron speaker coupled with an audio amplifier able to generate enough power to excite the acoustic frequencies in the range 50–500 Hz. The LMS® SCADAS III acquisition system was used to measure the acceleration levels. The Bruel & Kjaer® Pulse Data acquisition Unit type 3560 C was used for sound intensimetry. A series of high pass filters having a cut-off frequency of 6 Hz were introduced to filter unwanted perturbations measured by the microphones due to the measurement set-up. The PI® P-010.10P PICA Stack Power actuators were utilized for actively controlling the smart window prototype. The high voltage signal for the control of the piezoelectric actuators was provided by the PI® E-471 High Power Amplifier converting the inputs from the controller into the voltage for actuators. The active noise control system was implemented in a digital dSPACE® DS1103 PPC Controller Board with real-time processor and comprehensive I/O. The board was mounted on a Expansion Box to test the control functions in the laboratory set-up. The hardware is shown in Fig. 6.7. All the programs to acquire data and to implement controllers were developed in MATLAB® and Simulink® using Real-Time Workshop®.

6.4 Imperfections in the Experimental Set-Up

Active control of sound transmission through a multi-wall structure is tested effectively only if most of the sound energy is transmitted via the vibrations of the panels as well as through the cavities between them. If a fraction of

6.4. IMPERFECTIONS IN THE EXPERIMENTAL SET-UP

(a) LMS® SCADAS III R Data Acquisition system

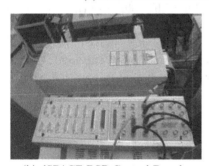

(b) dSPACE DSP Control Board

(c) PI® E-471 High Power Amplifier

(d) Harman Kardon Amplifier

Figure 6.7. Instrumentation

198   CHAPTER 6. ACTIVE NOISE CONTROL EXPERIMENTS

that energy is transmitted via peripheral paths, due to air-borne and/or structure-borne flanking transmission, the performance of the active noise control system can dramatically degrade. For these reasons, it has been necessary to analyze the vibro-acoustic response of the sending room by means of numerical and experimental investigations.

6.4.1 *Coupled vibro-acoustic model of the sending room*

A sketch of the model under study is shown in Fig. 6.8. The set-up comprises the soundproof box and the smart window test-bed mounted on the side opening of the box. The discretized test structure is shown in Fig. 6.9. The box is made of steel with a Young's modulus E, of 210 GPa; a Poisson's ratio ν of 0.3 and a density ρ of 7850 kg/m^3. An ambient density $\rho_0 = 1.225$ kg/m^3, a bulk modulus of 0.141 MPa and a speed of sound $c = 340$ m/s are used as fluid properties for the air in the cavities. The box is meshed with 4315 linear quadrilateral shell elements. First-order hexahedral acoustic elements (AC3D8) are used to fill in the volume of the interior box. A surface-based interface between the interior air and the box is modelled in ABAQUS® by using the *TIE option. This results in a coupling between the motion of the acoustic medium and the motion of the solid. One side of the box is partially open and the other four boundaries of the cavity are

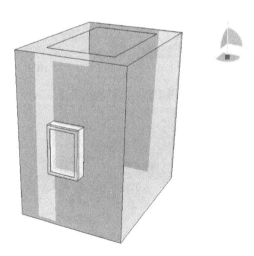

Figure 6.8. Model of the vibro-acoustic facility

6.4. IMPERFECTIONS IN THE EXPERIMENTAL SET-UP

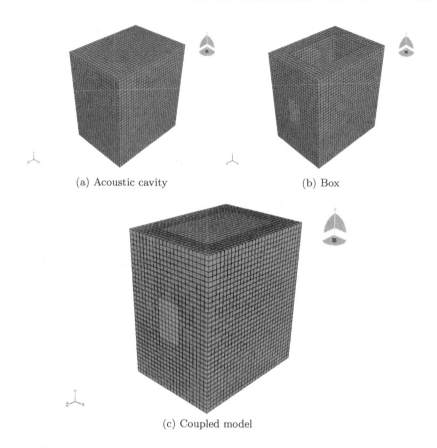

(a) Acoustic cavity

(b) Box

(c) Coupled model

Figure 6.9. Finite element discretization of the box and the interior cavity

acoustically coupled with the structural surfaces. The effect of interacting air at the outside of the box is neglected.

Natural frequency analysis

The results of the natural frequency analysis of the coupled vibro-acoustic system including the window prototype are illustrated in Table 6.1. The numerical predictions are also compared with the experimental results described in Section 6.4.2. The influences of the uncoupled acoustic and structural modes are evaluated. The analysis indicates that the effect of the interior acoustic pressure modelled inside the box is almost negligible. There is a good correspondence between the resonance frequencies predicted by the uncoupled and coupled model in the low frequency range.

Table 6.1 Comparison between analytical, numerical and experimental vibro-acoustic modes of the testing facility for sound transmission characterization

	Analytical		Numerical			
	Cavity (Hz)	Box (Hz)	Cavity–Box (Hz)	Cavity–Box–Window (Hz)	Experimental (Hz)	Error (%)
(0,1,0)	141.67	—	141.68	141.78	151.35	6.32
	—	143.56	143.54	143.72	—	—
	—	153.97	153.79	154.18	—	—
(0,0,1)	160.37	—	160.36	160.19	171.80	6.75
	—	—	—	191.73	197.00	2.67
(0,1,1)	213.98	219.09	213.84	213.95	225.34	5.05
(1,0,0)	215.19	—	215.36	215.51	227.81	5.4

Due to the influence of the added fluid mass, the natural eigenfrequencies of the coupled model are shifted towards lower values. A good numerical-experimental correlation is achieved and the first eigenfrequency of the window prototype is well approximated. The vibro-acoustic mode shapes are illustrated in Figs. 6.10 and 6.11. The former describes the pressure distribution in the acoustic cavity as well as the structural displacement. The latter illustrates the coupled responses. A frequency sweep in the vicinity of the first eigenfrequency of the window prototype is simulated by applying a planar sound wave generated in the sending room and directed towards the triple-panel partition. The frequency dependent planar incident wave, moving in the z-direction of the box, is modelled in ABAQUS® and the resulting vibro-acoustic response is then calculated. In such a fully coupled problem, both the sending room and the triple-panel partition are coupled with their respective enclosed air. This pairing creates an internal coupling between the acoustic pressure and structural displacements at the acoustic-structure interfaces. Figure 6.12 shows the pressure distribution obtained at 189.2 Hz for a given z-directed plane wave. A section of the reference box is plotted. The incident wave propagates in the positive z-direction and it is of unitary amplitude. The direction is defined by an appropriate location of the source and standoff points defined in ABAQUS®. This frequency is close to the first vibro-acoustic resonance of the window prototype. The resulting response is therefore dominated by the (1,1) mode shape. This mode is expected to exhibit a very efficient coupling to the radiated sound field due to the high sound radiation efficiency.

6.4. IMPERFECTIONS IN THE EXPERIMENTAL SET-UP

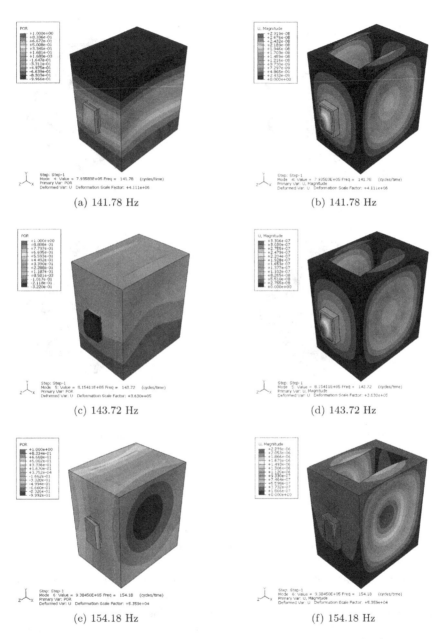

(a) 141.78 Hz (b) 141.78 Hz
(c) 143.72 Hz (d) 143.72 Hz
(e) 154.18 Hz (f) 154.18 Hz

Figure 6.10. Vibro-acoustic mode shapes of the testing facility: pressure (left), displacement (right), PART I

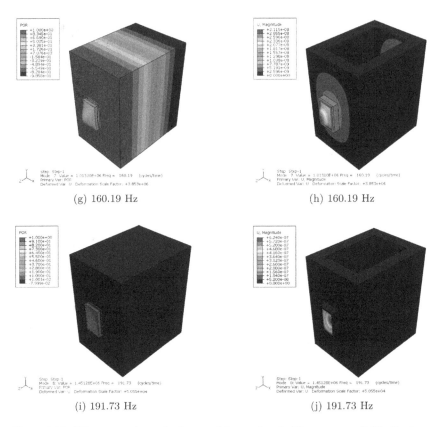

(g) 160.19 Hz   (h) 160.19 Hz

(i) 191.73 Hz   (j) 191.73 Hz

Figure 6.10. Vibro-acoustic mode shapes of the testing facility: pressure (left), displacement (right), PART II

6.4.2 *Experimental validation*

The vibro-acoustic simulation model of the sending room was validated through a series of experiments. Both the frequency response functions associated with the sound pressure in the cavity and the structural response of the panels were measured, as shown in Fig. 6.13. The interior acoustics of the box was characterized by exciting the acoustic enclosure via a speaker. The structural vibrations were evaluated by driving the piezoelectric actuators mounted on the smart window prototype. The input voltage signals to the loudspeaker as well as to the piezoelectric actuators were used as reference for the FRFs. A band-limited input signal up to 300 Hz was considered.

Figures 6.14 and 6.15 show the acoustical and the vibrational FRFs, respectively. The curves in Fig. 6.14 show four frequency-response functions

6.4. IMPERFECTIONS IN THE EXPERIMENTAL SET-UP

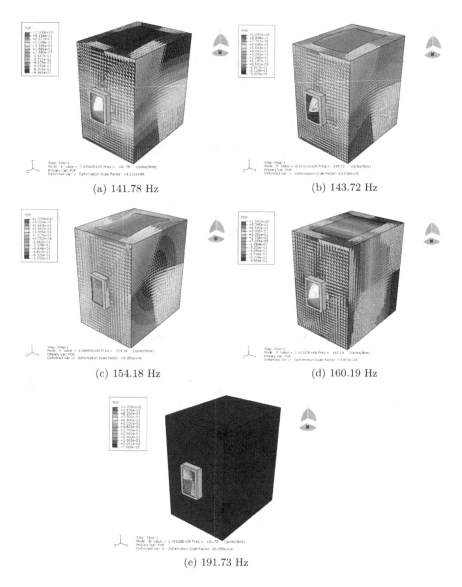

(a) 141.78 Hz

(b) 143.72 Hz

(c) 154.18 Hz

(d) 160.19 Hz

(e) 191.73 Hz

Figure 6.11. Coupled mode shapes of the vibro-acoustic facility

in terms of acoustic pressure in the acoustic cavity per voltage applied to the piezoelectric actuators (top figure) and to the internal loudspeaker (bottom figure) obtained with frequency sweeps from 100 to 300 Hz. The measurement points are illustrated in Fig. 6.13b. The curves in Fig. 6.15

204   CHAPTER 6. ACTIVE NOISE CONTROL EXPERIMENTS

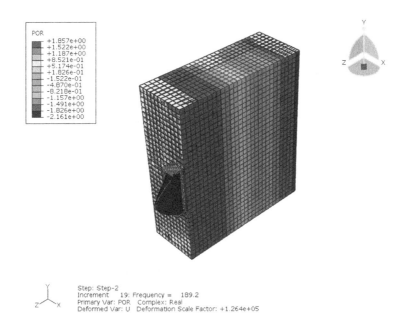

Figure 6.12. Fluid displacement for a plane wave excitation

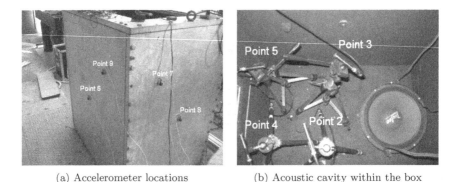

(a) Accelerometer locations   (b) Acoustic cavity within the box

Figure 6.13. Experimental characterization of the sending room dynamics

show four frequency-response functions in terms of structural acceleration in the box per voltage applied to the piezoelectric actuators (top figure) and to the internal loudspeaker (bottom figure). The structural point sensors are shown in Fig. 6.13a.

The acoustic field in the enclosure shows a dominant vibro-acoustic mode at 151.35 Hz. This mode is influenced by the uncoupled acoustic

6.4. IMPERFECTIONS IN THE EXPERIMENTAL SET-UP

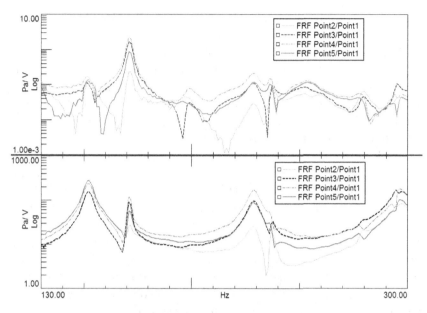

Figure 6.14. Noise spectra within the enclosure due to the actuation of the smart window prototype (top) and due to an interior acoustic excitation (bottom)

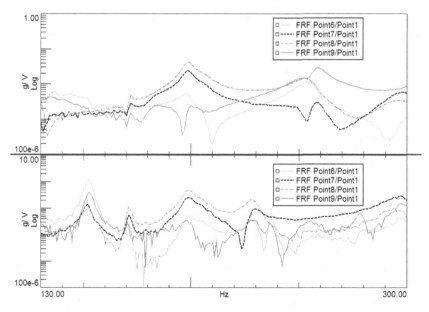

Figure 6.15. Vibrational spectra of the sending room due to the actuation of the smart window prototype (top) and due to an interior acoustic excitation (bottom)

mode of the box interior and, consequently, it is well excited by the acoustic source, as illustrated in Fig. 6.14 (bottom figure). Such a resonance peak also has an effect on the structural response of the box, as shown in Fig. 6.15 (bottom figure). In addition, a further peak in the noise spectra can be observed at 171.80 Hz. This frequency dominates the acoustic responses to both structural and acoustic excitation, as shown in Fig. 6.14 (top figure) and in Fig. 6.14 (bottom figure), respectively. This mode can be attributed to the (0,0,1) mode of the acoustic enclosure.

The low-frequency structural behaviour of the reference box is scarcely influenced by the actuation of the smart window prototype. The FRFs in Fig. 6.15 show a negligible effect at 197 Hz, measured only by Point 7 and Point 8, as shown in the top curves of Fig. 6.15. This frequency is very close to the first coupled resonance of the window prototype. This means that a small amount of vibrational energy is transmitted from the window to the entire box when the piezoelectric actuators are activated at that frequency. Above 200 Hz, the coupled system exhibits two acoustically dominated responses at around 230 Hz measured by the microphones installed within the box. They are the (0,1,1) and (1,0,0) acoustic modes, respectively.

Due to the dynamics of the sending room, experiments reveal that the laboratory set-up may amplify some frequencies more than others. Although these effects are less remarkable in the frequency range up to 150 Hz, they may affect the experimental results of active sound transmission attenuation through the smart window. Therefore, in order to better characterize the sound radiation emanating from peripheral paths, the window area as well as the entire box will be scanned with a sound intensity probe during the active noise control experiments detailed in Section 6.6.

6.5 Real-Time DSP Implementation of Active Noise Control

This section discusses the experimental investigation of the active reduction of sound transmission through the designed smart aircraft window set-up, shown in Fig. 6.5. The implementation of the filtered-reference feedforward algorithm on a real-time control system is also presented. The piezoelectric actuators mounted on the smart panels are driven in order to reduce the radiated sound field by transmitting the control forces to the radiating panel. The acoustic excitation is obtained by driving the primary sound source within the enclosure. The reference signal for the feedforward algorithm is the driving signal of the loudspeaker in the enclosure. This choice is

6.5. REAL-TIME DSP IMPLEMENTATION

representative of the actual situation in a propeller aircraft where the tacho signal from the engine is taken as reference. In some preliminary tests, not included in this book, one of the microphones in the box was also used as reference. However, in this case the performance of the active control system degraded.

Tonal and narrowband acoustic excitations are considered. As the real excitation spectrum for a turboprop aircraft was not available, an artificial harmonic disturbance was shaped to give a similar interior noise spectrum to that inside a turboprop aircraft in cruise conditions, such as the one presented in Fig. 1.5. The fundamental frequency was chosen to equal 124 Hz and the excitation signal contained harmonics up to 2 kHz. For the narrowband excitation, a stepped sine wave signal ranging from 120 to 130 Hz in ten seconds was investigated. This condition is analogous to the situation in a propeller aircraft where the frequency content of the acoustic excitation changes during flight due to a slightly varying engine rotational speed. A steady-state sinuisoidal disturbance was also selected in correspondence to the local minimum in transmission loss predicted by the numerical simulations.

6.5.1 The real-time controller

An adaptive real-time feedforward controller was developed for the proof-of-concept of active noise control. The controller prevents the noise from passing from the sending room to the anechoic chamber via the smart window. Compared to the numerical control system described in Chapter 4, the primary and secondary disturbances as well as the resulting error and monitoring signals were replaced by input and output channels for the DSP control board connected to the concerned physical systems, i.e. the loudspeaker, the active window prototype and the control microphones, respectively. Moreover, the noise passing through the window was monitored by further three microphones connected to the DSP board. The adaptive control system was realized with digital filters whose coefficients were iteratively adapted by processing the measured error signals. The control loop also included digital low-pass digital filters to eliminate high frequency background noise captured by the microphones.

A SISO FXLMS controller was developed and tested first. This means that only one microphone was used as error sensor and all the piezo actuators were driven in parallel. The code was developed in the graphical development environment of MATLAB/Simulink®. The C code was generated in Real-Time Workshop® of MathWorks® and then implemented in the

208   CHAPTER 6.   ACTIVE NOISE CONTROL EXPERIMENTS

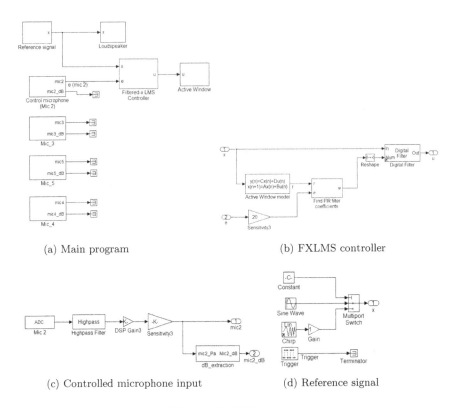

Figure 6.16. SISO code

dSPACE® DS1103 PPC Controller Board. Some of the block diagrams of the developed real-time control system are illustrated in Fig. 6.16. The main program is shown in Fig. 6.16(a). The core of the program is the *Filtered-x LMS Controller* block, which takes as inputs the *Reference signal* and the *Control Microphone*. It then feeds the active window through the piezoelectric actuators to minimize the *Control Microphone* signal. For a SISO approach, the control actuators of the smart window are driven in parallel. The operating voltage of the piezoelectric actuators is limited within practical limits represented by ±400 V. This protection allows the system to directly switch off the signal feeding the active window if it overcomes the prescribed voltage. The *Reference signal* also drives the *Loudspeaker*, producing the disturbance signal to be controlled. Figure 6.16(b) shows the logic of the FXLMS controller. This block contains the feedforward approach and includes the experimental plant model of the active window used to filter the *Reference signal*. The plant model is reconstructed from

6.5. REAL-TIME DSP IMPLEMENTATION

the knowledge of the transfer function of the secondary path measured in previous experiments. It describes the complete vibro-acoustic behaviour of the window prototype driven by the actuators. The instantaneous noise residue, measured by the *Control Microphone*, is utilized to adjust the adaptive filter coefficients in an attempt to minimize the transmitted noise. The FIR filter coefficients are computed inside a masked block which includes the convergence coefficient α and the FIR filter length as tunable parameters. The *Control Microphone* block, shown in Fig. 6.16(c), basically scales the measured signal in order to derive the corresponding pressure signal. Furthermore, the program processes the error signal in order to perform a real-time computation of the sound pressure levels, expressed in dB. The *Reference signal* block, shown in Fig. 6.16(d), allows the user to switch the fundamental frequency of the acoustic disturbance between a sine wave for a given Blade Passage Frequency (BPF) and a chirp signal for a slightly varying BPF.

Three graphical interfaces were developed in dSPACE ControlDesk in order to manage the real-time signals measured during the experiments. The real-time monitoring codes developed for a SISO application are shown in Figs. 6.17–6.19. Such graphical-user interfaces were developed to monitor the DSP output and input signals as well as the control performance and stability. The first panel allows the user to command the state of the

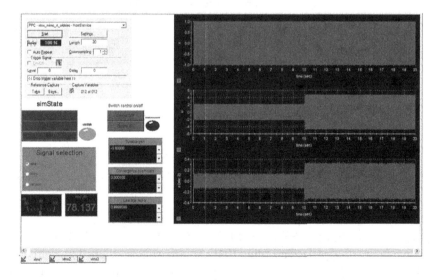

Figure 6.17. Control Panel noise attenuation measured by the control microphone

210   CHAPTER 6. ACTIVE NOISE CONTROL EXPERIMENTS

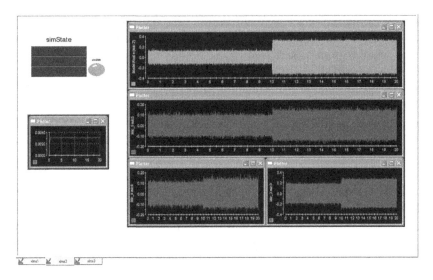

Figure 6.18. Monitoring Panel noise attenuation measured by the control and monitoring microphones

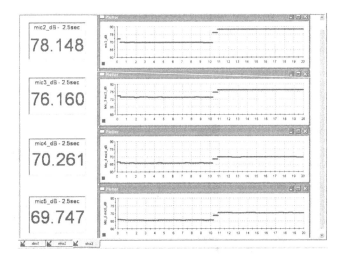

Figure 6.19. Real-time computation of dB values

program running the DSP control board. A LED indicates the state. This state can be *stop*, *pause* (meaning that the program is stopped but already initialized) or *start*. In this panel, the user can set a number of parameters, such as the duration of the acquisition, the sample time, the triggering,

6.5. REAL-TIME DSP IMPLEMENTATION

the path of the file to save or the name of the file. Three different types of acoustic primary disturbance can be chosen. A pushbutton can switch the controller on or off. Some of the controller parameters can also be modified, such as the convergence coefficient α.

The error signal, as well as the acoustic disturbance, are monitored on the right part of the Control Panel, shown in Fig. 6.17. From the top, the time histories of the reference signal, the input voltage to the piezo actuators and the sound pressure measured by the error microphone are plotted. In the Monitoring Panel, shown in Fig. 6.18, the user displays the sound pressure signals measured by the control microphone together with those acquired by the three monitoring microphones. They are time histories of the sound pressure plotted in Pascal. In the third panel, shown in Fig. 6.19, a real-time computation of the noise pressure level expressed in dB is performed in order to provide the user with the current estimate of the sound attenuation achieved by the smart window controller.

Before doing the experiments, the sound pressure levels acquired by the developed control system were validated. The accuracy of the amplitude and phase measured by each microphone was checked with a Sound Calibrator, B&K Type 42.31, giving a sinusoidal signal of 94 dB at 1 kHz, as shown in Fig. 6.20. Such a test is shown in Fig. 6.20c. As shown in the control panel illustrated in Fig. 6.21, there is a good agreement between the sound pressure level measured by the real-time control system (93.955 dB) and the nominal acoustic excitation of 94 dB applied through the piston-phone.

6.5.2 *Experimental results*

The control system capabilities in adaptively attenuating the acoustic field transmitted through the window were assessed experimentally. By assuming that the off-line identified plant response of the window prototype was accurate enough to represent the behaviour of the secondary paths, the real-time levels of noise were measured and compared with and without control. Experiments were conducted for off-resonance cases first and sound attenuation performance was investigated by increasing the number of control actuators. This is because in resonance conditions, the acoustic response is mainly dominated by only one vibro-acoustic mode of the structure. In general, as the structural response is dominated by multiple modes, a proportional number of actuators is needed to provide the control input unless these modes are inefficient acoustic radiators. Then, the loudspeaker was

212 CHAPTER 6. ACTIVE NOISE CONTROL EXPERIMENTS

(a) Sound Calibrator, B&K Type 42.31

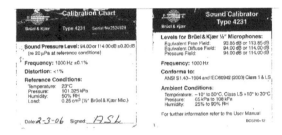
(b) Calibration chart

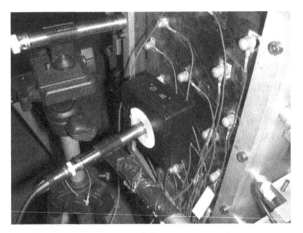
(c) Sound calibration of the error microphone

Figure 6.20. Microphone calibration

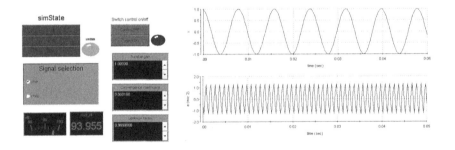

Figure 6.21. Sound pressure level computation [dB] from noise signals in Pascal

6.5. REAL-TIME DSP IMPLEMENTATION

driven at the first mass–air–mass resonance of the triple-wall system, which was experimentally controlled only for a SISO control configuration.

Before applying the control, the acoustic field transmitted through the structure, as well as the structural response of the radiating panel to the acoustic excitation, were measured. The excitation level at the incident panel was 120 dB, resulting in a transmitted noise in the anechoic room of approximately 80 dB (off-resonance). This level is within an order of magnitude for an actual aircraft cabin. After obtaining measurements for the uncontrolled case, the filtered-X LMS algorithm was invoked and the radiating panel of the triple-wall structure was controlled via the piezoelectric actuators. The control parameters were adjusted to obtain the best control performance in combination with a reasonable control effort while maintaining system's stability. Upon converging to the minimum acoustic response at the control microphone, acoustic and structural measurements were performed. After obtaining these measurements, the uncontrolled and controlled cases were compared for both the acoustic radiation and the structural vibrations. Of the piezoelectric actuators mounted on the smart window prototype, some were connected in parallel and controlled by one channel of the controller. In the basic SISO configuration, the piezo actuators were grouped together and driven in phase. In the multi-channel active control system, the "square" case of four error sensors and four control actuators was considered. This is because when the number of secondary sources is equal to the number of error sensors, the resulting square matrix of transfer functions is well conditioned, thus providing the best performance at the control points. In Fig. 6.22, such combinations are shown. The actuator and microphone locations are shown in schematics. Figure 6.23 shows the test

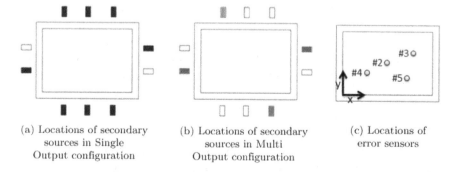

(a) Locations of secondary sources in Single Output configuration

(b) Locations of secondary sources in Multi Output configuration

(c) Locations of error sensors

Figure 6.22. Location of control actuators and error sensors

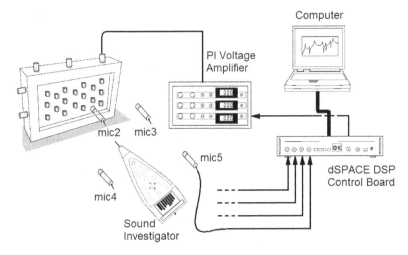

Figure 6.23. Schematic of the active noise control experiments with the feedforward controller connected

Table 6.2 Location of error sensors

Sensor	x (mm)	y (mm)	z (mm)
Mic. 2	160	100	100
Mic. 3	240	130	150
Mic. 4	80	60	700
Mic. 5	200	50	1050

arrangement in the anechoic chamber. The control and monitoring microphones used throughout the experiments are positioned in the receiving room with an offset of 100 mm, 150 mm, 700 mm and 1050 mm from the surface of the radiating panel. The locations of the error sensors are given in Table 6.2. The primary noise source inside the enclosure was driven by combining the DSP-based digital signal generator with the audio amplifier. The control signals from the controller were power amplified into the voltage for the actuators.

Single output control

This section describes the experimental results achieved by implementing the filtered-X LMS algorithm in a DSP control board. A SISO controller is investigated first by considering one error sensor and by driving all the

6.5. REAL-TIME DSP IMPLEMENTATION

piezoelectric actuators in phase. The remaining acoustic sensors are not used in the control loop but they are available to measure the sound field attenuation levels with and without control. Next, the sound attenuation is investigated by increasing the number of error sensors included in the control loop.

One error sensor (1i1o)

One error sensor, three monitoring sensors, eight actuators and one reference signal were used for SISO experiments. This means that only one error microphone was used as input to the control algorithm while eight actuators were grouped and driven together by one channel of the controller. Figure 6.24 shows a picture of the tests executed. Figure 6.25 shows the measured sound pressure signal at the error microphone (mic. 2) with and without the single channel feedforward controller. Its left part represents the resulting sound pressure when the control is turned ON, while its right part (after approximately ten seconds) is the resulting sound pressure when the control is turned OFF. Such data are obtained by post-processing the real-time active control results measured by the data acquisition codes illustrated in Figs. 6.18 and 6.19. The results of the single-channel controller are also plotted in Fig. 6.26 in the respective third-octave bands with and without control. No spill-over effects (unintentional excitation of higher

Figure 6.24. Validation of the SISO active noise control results by using a B&K Sound Investigator

216 CHAPTER 6. ACTIVE NOISE CONTROL EXPERIMENTS

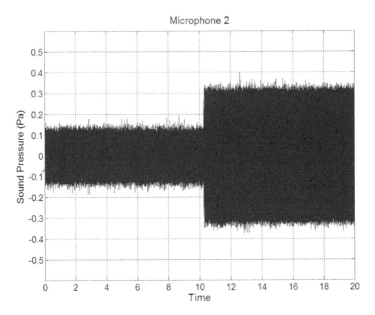

Figure 6.25. Noise attenuation at the control microphone (mic. 2) in a SISO configuration (control OFF after approximately ten seconds)

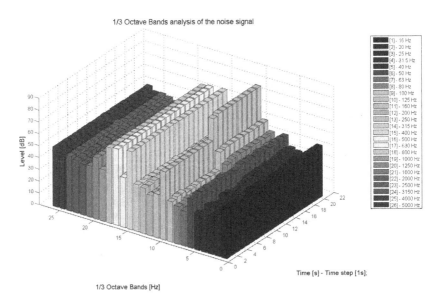

Figure 6.26. Measured third-octave levels with and without control for the SISO configuration

6.5. REAL-TIME DSP IMPLEMENTATION

harmonic frequencies) are observed. Without control the maximum level is found to be in the 125 Hz third-octave band corresponding to the fundamental frequency of the acoustic excitation. The obtained level reduction at that frequency equals 15 dB.

In order to validate the results obtained for the active control of sound transmitted through the smart window prototype, the controlled sound field was also separately measured by using a calibrated Sound Investigator B&K 2260. The main results obtained with the ASAC system are summarized in Table 6.3. They are given in terms of the sound pressure levels measured in the receiving room by the error microphone as well as by the monitoring microphones with and without control. As shown in Fig. 6.27, the experimental results match very well the values expressed

Table 6.3 Measurement results for the SISO configuration

	Radiated sound pressure [dB]		
	Control OFF	Control ON	Noise reduction
Control mic. 2	78.22	70.43	7.79
Monitoring mic. 3	76.05	72.00	4.05
Monitoring mic. 4	70.33	67.32	3.01
Monitoring mic. 5	69.95	65.68	4.27

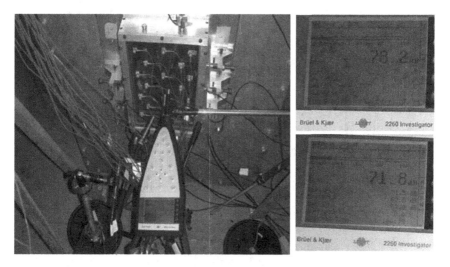

Figure 6.27. Sound Investigator measuring the sound transmission attenuation at the control microphone (mic. 2), (above) Control OFF, (below) Control ON

by the Sound Investigator positioned close to the control microphone. The maximum attenuation, achieved in correspondence to the control microphone, amounts to around 7.8 dB. It can be noted that the sound pressure level is also attenuated at the monitoring sensors, although they are not included in the control loop. Their sound attenuation ranges between 3 and 4 dB.

For the on-resonance case, a steady state sinusoid at 197 Hz was used as disturbance signal. In this experiment, the acoustic excitation was tuned for a resonant response of the window structure corresponding to the mass–air–mass resonance frequency. This eigenfrequency is characterized by a strong coupling between the structure and the enclosed fluid domain. Since the response at that frequency is dominated by the uncoupled structural (1,1) mode of the radiating panel, a SISO controller was sufficient to reduce the sound radiation effectively. Results of active control of the resonant noise transmission are presented in Fig. 6.28. Sound attenuation at the noise frequency (197 Hz) in the order of 20 dB was obtained through the modal reduction of such an efficient acoustic radiator.

Finally, the signal driving the control actuators was measured in order to evaluate the control effort. The actuator voltage measured during control

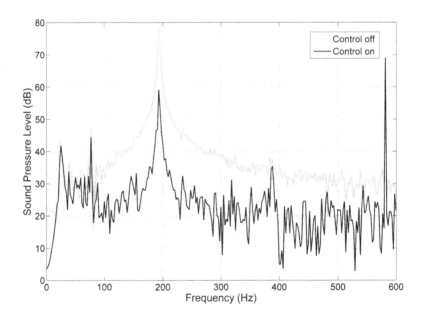

Figure 6.28. Sound attenuation at 197 Hz

6.5. REAL-TIME DSP IMPLEMENTATION

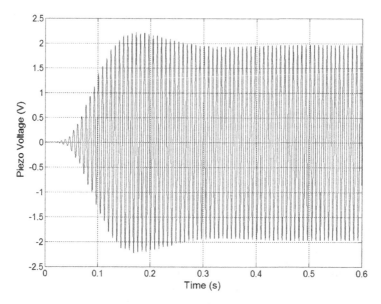

Figure 6.29. Actuator voltage applied to the piezoelectric actuators for a SISO control of sound transmission

is shown in Fig. 6.29. These values are the outputs of the controller which are then supplied to the respective piezo amplifiers providing a frequency constant gain of 100. The voltage signal is dominated by the fundamental frequency and a maximum voltage of 400 V peak-to-peak is applied to each actuator. In the SISO control configuration, the piezo actuators were driven in phase with an actuator workload, defined as the ratio between the actual maximum peak voltage and the absolute maximum voltage which can be applied to the actuator, less than 50%.

Four error sensors (4i1o)

Using all four microphones as error sensors slightly improved the active control of the sound transmitted through the window prototype. Experimental results for a MISO filtered-X LMS controller are illustrated in Table 6.4. This configuration consisted of four error inputs and one output (4i1o) driving eight piezoelectric actuators in parallel. A reduction in the overall radiated sound field was achieved between 4 and 8.5 dB with little discernible differences respect to the SISO configuration due to the increased number of error sensors. Little if any improvement in sound attenuation was observed at the error microphone 2. Such improvements were somewhat higher at

Table 6.4 Measurement results for the MISO configuration

	Radiated sound pressure [dB]		
	Control OFF	Control ON	Noise reduction
Control mic. 2	78.14	69.62	8.52
Control mic. 3	76.16	71.24	4.92
Control mic. 4	70.26	65.53	4.73
Control mic. 5	69.74	65.57	4.17

microphones 3, 4 and 5. A more uniform attenuation was obtained since all error sensors contributed to the computation of the secondary output signals driving the piezoelectric actuators.

Multi-output control

The experimental results presented in the previous section refer to a Single Output real-time controller driving all piezo actuators equally. In this section, the sound transmission control is evaluated by increasing the number of control actuators involved in the control loop. In this case, the phase and amplitude of each actuator can be varied more effectively with respect to the disturbance signal in order to attenuate the radiated noise field. This results in a more accurate and distributed control of the radiating panel.

Two error sensors (2i2o)

With respect to the previous configuration, a significant improvement in sound reduction was observed by increasing the number of control outputs. Among the possible combinations using two control output arrangements together with two error microphones, a multi-output configuration using the actuators nos 2, 5, 7 and 10 demonstrated enhanced performance. Only two error signals (mic. 2 and mic. 3) were minimized by the real-time controller while microphones 4 and 5 were only used to monitor the sound field attenuation. Morevoer, the actuators 5 and 7 at the top right corner were connected in parallel as well as the actuators 2 and 10 at the bottom left corner.

In general, multiple modes participate in off-resonance structural response. As the number of control channels increased, a more accurate phasing of actuators resulted in a more effective control of the acoustic radiators. While 8.5 dB of reduction in the radiated sound field was achieved at the control microphone when implementing a single channel system, almost 15 dB of sound attenuation was obtained when utilizing four control actuators. The results achieved for the multi-channel controller

6.5. REAL-TIME DSP IMPLEMENTATION

Table 6.5 Measurement results for the (2i2o) MIMO configuration

	Radiated sound pressure [dB]		
	Control OFF	Control ON	Noise reduction
Control mic. 2	82.00	67.16	14.84
Control mic. 3	78.91	68.45	10.46
Monitoring mic. 4	68.69	65.30	3.39
Monitoring mic. 5	68.29	66.35	1.94

are listed in Table 6.5. The better performance compared with the single channel controller is explained by the higher control authority provided by the control system driving the smart window prototype. A better sound attenuation was obtained by using optimized control signals in regard to their correlation to the radiated sound field.

In this configuration, the highest reduction in the sound transmission was obtained at microphone 2. This result is not surprising since the error criterion is the sum of the squares of the microphone error signals. Since this microphone is the closest to the radiating panel, it has a major impact on the minimization of the cost function. The radiated sound field measured at that position was reduced by 15 dB overall. On the other hand, a very poor attenuation was achieved at the monitoring microphone 5 where the sound field was attenuated by only 2 dB. The time histories of the measured pressure signals are shown in Fig. 6.30. The left part of the signal represents the resulting sound pressure when the control is turned ON, while the right part (after approximately 18 seconds) is the resulting sound pressure when the control is turned OFF. Figure 6.31 shows the comparison between the frequency spectra of the noise signal measured by the control microphone 2 with and without control. At the fundamental frequency of the acoustic excitation corresponding to 124 Hz, the radiated sound field is well correlated to the disturbance. It was attenuated by the active control from 82 dB to 57 dB corresponding to a reduction of about 25 dB. For the harmonics at 248, 372 and 496 Hz, the decrease in radiated sound field amounts to 9, 8 and 4 dB, respectively.

Four error sensors (4i4o)

In most applications, a uniform zone of quiet is desired since differences in sound levels can be easily perceived by passengers as a potential cause of discomfort. The use of a single error control system, in fact, results in a residual acoustic field having significant differences between the minimum

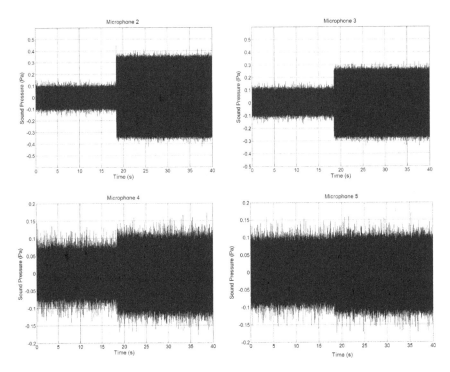

Figure 6.30. Time histories of the measured sound pressure signals with and without control in a (2i2o) MIMO configuration (control OFF, after approximately 18 seconds)

and maximum sound levels measured by the control and monitoring microphones. By using a multiple input controller, the noise reduction distribution can significantly improve.

Driving a number of actuators with a different phase and amplitude contributes to improving the control authority of the radiating panel. However, the number of control actuators and sensors of a real-time control system is limited by the processing power in real-time DSP implementations. Moreover, from a control point of view, it is essential to use an algorithm with a high convergence rate, good tracking performance and robustness. These qualities are often associated with increased computational complexity.

In order to reach a trade-off between the filtered-X operations of the LMS algorithm, requiring a number of computations that depend on the system complexity, and the control performance, which is expected to improve when the number of actuators and sensors is increased, the controller was implemented in a multi-channel configuration using four

6.5. REAL-TIME DSP IMPLEMENTATION

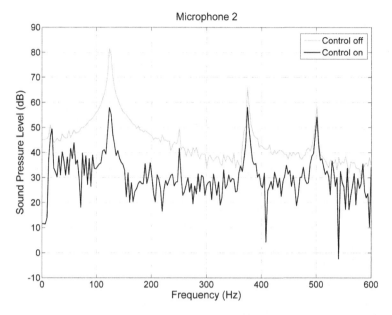

Figure 6.31. Comparison between the frequency spectra of the control microphone with and without a (2i2o) MIMO control

individual control outputs. Such a fixed number of control actuators reduces the hardware complexity and it can be more easily integrated into an aircraft sidewall panel due to the reduced costs involved.

The main block running the codes developed for the time-domain implementation of the MIMO filtered-X LMS controller is schematically illustrated in Fig. 6.32. Various parameters can be selected by the user via a real-time monitoring software running on the acquisition PC. The control filter coefficients of the multi-channel algorithm are computed iteratively by the steepest descent algorithm using a stochastic gradient. The reference signal is filtered by approximated models of the secondary paths and multiplied by the error signals measured at the nth sample time. The FIR filter coefficients are then adapted by implementing the MELMS algorithm, described in Eq. (4.48), to provide the input signals to the piezoelectric actuators.

The developed real-time control system was tested by increasing the number of control outputs from two to four and by using four error microphones at the same coordinates as in the previous tests. As shown in Fig. 6.22, the actuators 2, 5, 7 and 10 were selected for the control loop.

224 CHAPTER 6. ACTIVE NOISE CONTROL EXPERIMENTS

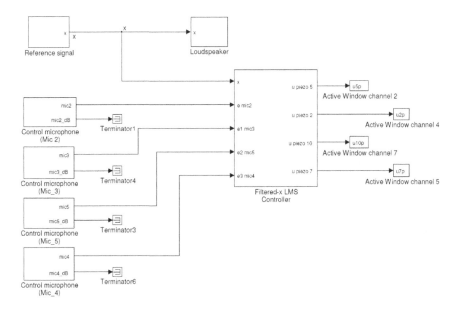

Figure 6.32. Main block of the MIMO code

Such actuators proved to be capable of controlling the most efficient acoustic radiators. Figure 6.33 shows the controller success at each of the error microphones in terms of sound pressure measured with and without control for a harmonic excitation. The left part of the signal represents the sound pressure measured when the control is turned ON, while the right part (after approximately 20 seconds) is the sound pressure when the control is turned OFF. The results achieved with the control of harmonic disturbance are summarized in Table 6.6. In contrast to the results obtained by the (2i2o) experiments, minor sound attenuation was observed at the error microphones 2 and 3. The error microphone 4 showed the same basic noise reduction. Conversely, signal residue of microphone 5 was reduced and it was comparable to that of microphone 4. As observed for the single output controller, increasing the number of error sensors resulted in a more uniform spatial distribution of sound attenuation and, thus, in a larger quiet area around the error microphones. This is an important feature to be considered as a more uniform useful-size zone of quiet is essential for the head motion of a person seated next to the aircraft window.

As shown in Fig. 6.34, the frequency spectrum computed at microphone 2 comprises four dominant frequencies: the fundamental frequencies, i.e. the frequency of the disturbance radiated from the propeller, and three

6.5. REAL-TIME DSP IMPLEMENTATION

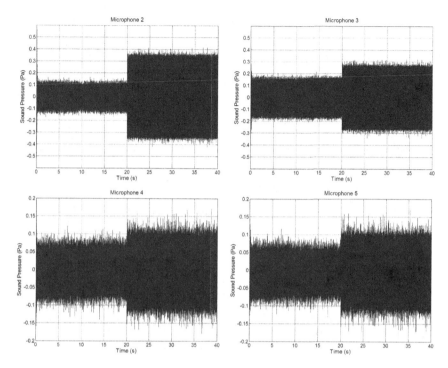

Figure 6.33. Time histories of the measured sound pressure signals with and without control in a (4i4o) MIMO configuration (control OFF, after approximately 20 seconds)

Table 6.6 Measurement results for the (4i4o) MIMO configuration

	Radiated sound pressure [dB]		
	Control OFF	Control ON	Noise reduction
Control mic. 2	81.40	68.47	12.93
Control mic. 3	79.01	71.75	7.26
Control mic. 4	68.53	64.72	3.81
Control mic. 5	67.28	63.20	4.08

harmonics. Noise attenuation at 124 Hz reaches 23 dB when the controller is switched on. On the contrary, an increase in the sound radiation is observed for the second and third harmonics of about 10 dB and 4 dB respectively with the closed control loop. Notice that the sound pressure levels are below 40 dB in the controlled conditions, except for harmonics associated with the noise disturbance. These values of sound pressure level are very close to the lowest threshold of human perception.

226   CHAPTER 6. ACTIVE NOISE CONTROL EXPERIMENTS

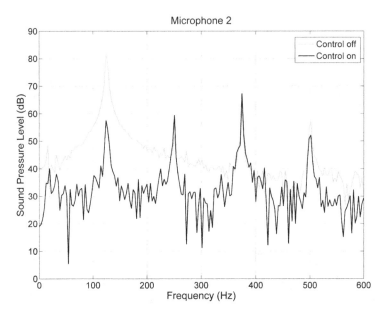

Figure 6.34. Comparison between the frequency spectra of the control microphone with and without a (4i4o) MIMO control

Moreover, the secondary signals induced by the actuators lead to a reduction in the structural acceleration levels measured on the radiating panel. They are attenuated although not included in the control loop. This is illustrated in Fig. 6.35, showing the time histories of the acceleration signals 8 and 9 for a period of two seconds and one-tenth of a second with and without control. Sensor 8 is located approximatively at the centre of the radiating panel, as shown in Fig. 5.28. Its vibration level is reduced by more than 50% with active control. On sensor 9, mounted closer to one of the control actuators, the vibration reduction reaches 70%. The corresponding vibration spectra are plotted in Fig. 6.36. At the fundamental frequency a reduction of about 17 dB is achieved. The vibration levels at the remaining harmonics are of the same order or slightly increased.

In many control applications, the robust stability is fundamental in order to maintain good performance in response to changes in the plant response as well as in the spectral components of the disturbance. For example, piezoelectric response may vary with time and operating conditions, due to temperature variations. In addition, digital controllers exhibit inherent delays associated with the antialiasing and reconstruction filters,

6.5. REAL-TIME DSP IMPLEMENTATION

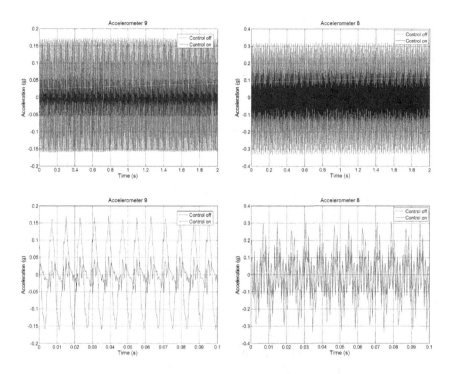

Figure 6.35. Acceleration measured by sensors 8 and 9 with and without control

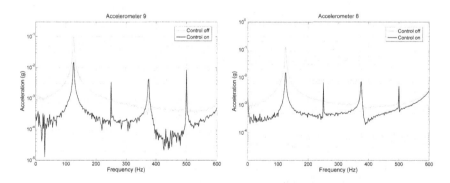

Figure 6.36. Acceleration spectra with and without control

as well as with the processing delays. Both phenomena may affect active noise control performance.

In an attempt to evaluate the controller adaptation under changing conditions, the real-time control system was tested against a non-stationary acoustic perturbation. Each plant response of the multi-channel controller was identified off-line by applying white noise to the secondary actuators in turn. They were assumed time invariant. A stepped-sine acoustic signal in the frequency range between 120 and 130 Hz was employed as acoustic disturbance. A bi-directional linear frequency sweep was used with a duration of 10 seconds. Although the plant response did not change, the controller was therefore tested to track changes in the acoustic excitation. The convergence coefficient α in the algorithm was adjusted until it was judged that the convergence rate was fastest. The real-time controller was able to track acoustic excitations varying in frequency from given initial conditions by rapidly changing the filter coefficients. The experimental results obtained for the narrowband excitation are shown in Fig. 6.37. The curves represent both the time histories and the spectograms computed for each error sensor processed by the MIMO controller. The former demonstrate a substantial increase in the sound pressure signals while turning the control system off (after about 22 seconds), especially at the error microphones positioned close to the smart window. A lower sound attenuation was achieved at the microphones 4 and 5. The latter illustrate the corresponding spectograms. The time evolution of the spectral content of the acoustic response was analyzed by dividing the non-stationary signals into a sequence of time segments where they could be treated reasonably as stationary. After that, a Short Time Fourier Transform was computed and results represented in a time-frequency plot. The time dependency of the sound attenuation levels, expressed in dB, with and without control, is shown in Fig. 6.38. They were measured every 0.5 seconds. Attenuation higher than 15 dB was achieved. Experiments reveal that active control performance is sensitive to the variation in the excitation frequency. The resulting sound attenuation is not constant but it degrades at some frequencies. Moreover, the same behaviour is observed for both increasing and decreasing frequencies.

As shown in Fig. 6.39, the radiated sound pressure attenuation was associated with a reduced structural response. This demonstrates that the real-time controller mitigates the acoustic radiation at the error sensors indirectly by reducing the vibration levels in the structure. Such a performance was observed for almost all the accelerometers bonded to the structure. The active control was turned off after approximately 17 seconds, resulting in a substantial increase in the residual vibrational field. The time dependency of the corresponding frequency spectrum is also provided. Vibration reductions of up to 12.5 dB were achieved.

6.5. REAL-TIME DSP IMPLEMENTATION

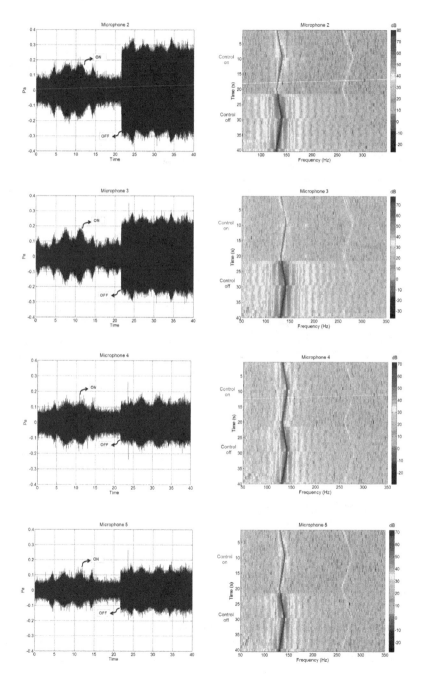

Figure 6.37. MIMO active noise control of narrowband disturbance (control OFF after approximately 22 seconds); spectograms with and without control (right)

230   CHAPTER 6.   ACTIVE NOISE CONTROL EXPERIMENTS

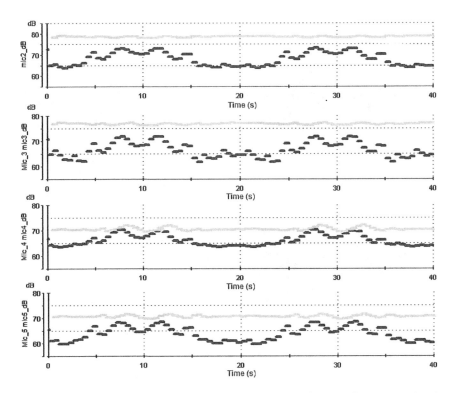

Figure 6.38. Real-time computation of sound attenuation (dB value) for narrowband acoustic excitation, control OFF (gray) vs. control ON (black)

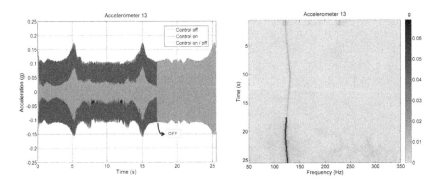

Figure 6.39. Comparison between the vibration levels measured with and without control for a narrowband acoustic disturbance (left); time frequency analysis of the acceleration levels (right)

6.6 Sound Intensity Measurements

The total radiated sound power is an attractive metric with which to evaluate active noise control performances. As mentioned in Section 2.3, it can be derived from the experimental evaluation of the sound intensity level. The relationship between sound intensity level and sound pressure level at any points of the sound field depends on the characteristics of the sound source and the measurement environment. The instantaneous sound intensity is defined as the rate of flow of sound energy per unit of surface area in the direction of local instantaneous acoustic particle velocity [180]. It is a vectorial quantity defined as the product of the sound pressure p and particle velocity $\mathbf{v}$:

$$\mathbf{I}(t) = p(t) \cdot \mathbf{v}(t). \qquad (6.1)$$

The time-averaged value of $\mathbf{I}(t)$ defines the Sound Intensity vector $\mathbf{I}$:

$$\mathbf{I} = \lim_{T \to \infty} \frac{1}{T} \int_0^T \mathbf{I}(t) \cdot dt \qquad (6.2)$$

where T is the integration period.

The integral over any surface totally enclosing the sound source of the scalar product of the sound intensity vector and the associated area provides a measure of the sound power Π radiated by all sources located within the enclosing surface. Therefore, the time-averaged rate of flow of sound energy through the area S of the measurement surface approximates the amount of sound power passing through the surface S. Notice that $\mathbf{v}$ is a vector quantity which may not be normal to that surface.

The sound power radiated by the window prototype into the anechoic chamber was determined by sound intensity measurements. This method is based on sampling the intensity field normal to the measurement surface by moving an intensity probe along one or more specified surfaces. In addition, the normal intensity components were measured over each surface of the vibro-acoustic facility in order to detect possible leaks and flanking paths in the sound transmission.

An intensity probe consists of two spaced microphones. A typical arrangement is depicted in Fig. 6.40. The microphones are separated by a spacer whose dimension depends on the upper frequency limit of measurement: the smaller the spacer, the higher the frequency that can be measured. In general, the acoustic wavelength must be six times the spacer distance. The low-frequency limit depends upon the phase mismatch of the microphone pairs, which is usually influenced by the acquisition hardware.

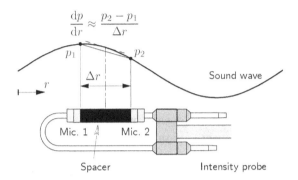

Figure 6.40. Estimation of the pressure gradient with a sound intensity probe [181]

The particle velocity is estimated at the acoustic centre of the probe, between the two microphones. In a linear sound field, the particle velocity and the pressure gradient are related by Euler's equation, which for propagation in one direction reads:

$$\rho_0 \frac{\partial v_r(t)}{\partial t} = -\frac{\partial p(t)}{\partial r} \quad (6.3)$$

where $v_r(t)$ is the particle velocity in the direction r. The velocity derivative with respect to time at the centre measurement point can be expressed by the numerical derivative of the pressure over the spacer length Δr. In the frequency domain, the particle velocity can be approximated as

$$v_r \simeq -\frac{1}{j\rho_0\omega} \frac{p_2 - p_1}{\Delta r} \quad (6.4)$$

where p_1 and p_2 are the pressure amplitudes measured by the microphones on the intensity probe. The pressure is instead approximated at the centre of the spacer by taking the average pressure of the two microphones, $p = (p_1 + p_2)/2$. The pressure and particle velocity signals are then multiplied together and time averaging gives the intensity. The sound power Π is then computed by multiplying the space average of the sound intensity vector by the surface area S.

Two different methods may be applied in order to measure the space-averaged sound intensity: the discrete point method and the scanning method. They are schematically illustrated in Fig. 6.41. The discrete point method consists of a number of discrete points where the probe is held for each measurement. The variability in the intensity over the measuring surface determines the number of points needed. If the variability is high the

6.6. SOUND INTENSITY MEASUREMENTS

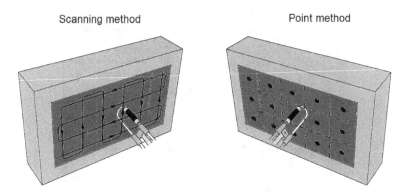

Figure 6.41. Sound power measurements with the scanning method and the discrete point method

number of measuring points must be increased. In the scanning method, the intensity probe is moved with constant speed across a path on the measurement surface. Both methods are very time consuming. As a result, the stability of the active control system can also be investigated during the sound power measurements.

6.6.1 *Experimental results*

Sound intensity measurements were carried out into the anechoic chamber with and without control in order to quantify the sound power attenuation achieved by the active noise controller driving the smart window. In addition, the noise transmission path through the sending box was detected and characterized. The spatial localization of the sources was obtained by the magnitude and direction of the energy flow over the individual surfaces. Regions with higher sound intensity were interpreted as locations of major sound transmission.

The acoustic excitation was provided by a loudspeaker located in the sending room. On the anechoic side, the window prototype as well as the entire reference facility were scanned by the B&K 3595 intensity probe with two half-inch microphones separated by a spacer of 50 mm. Most of the hardware in the experimental setup was already discussed in Section 6.3. The sound intensity probe as well as the data acquisition unit are shown in Fig. 6.42.

The discrete point method was used. Multiple flat surfaces enclosing the vibro-acoustic facility were defined with a regular measurement grid. The active and reactive intensity levels were measured by scanning manually on

234   CHAPTER 6. ACTIVE NOISE CONTROL EXPERIMENTS

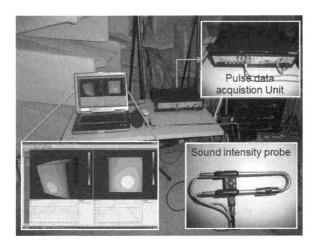

Figure 6.42. Instrumentation for sound intensity measurements

a grid of 52 points in a plane 500 mm distant from the window prototype. The measurement grid is shown in Fig. 6.43. Twelve measurement points were considered for each surface. The time of acquisition in each point was equal to ten seconds. The characteristic values of the set-up are $d = 500$ mm, $a = 2060$ mm and $b = c = 1800$ mm.

As the intensity analysis is particularly sensitive to the variation in magnitude and phase of the microphones, the two microphones of the intensity probe were amplitude calibrated first. The calibration was carried out with a special piston-phone providing a known intensity level.

The total sound power radiated by the window prototype into the anechoic chamber was measured by driving the internal source with an acoustic signal having the fundamental frequency of 125 Hz and harmonics up to 2 kHz superimposed with a band-filtered white noise signal. Figure 6.44 shows the sound intensity measured on the window area without the controller operation. A sound intensity peak of 67 dB was found in correspondence to the 125 Hz third-octave band and it was reached in the middle of the measurement grid. The local sound intensity map without control for the 125 Hz third-octave band is shown in Fig. 6.45. It was found that the window area as well as the upper part of the facility were the principal sound intensity paths. Results show that the sound energy transmitted via the window was about 6 dB higher than that transmitted through the walls of the box. This is sufficient as a qualitative indicator to experimentally

6.6. SOUND INTENSITY MEASUREMENTS

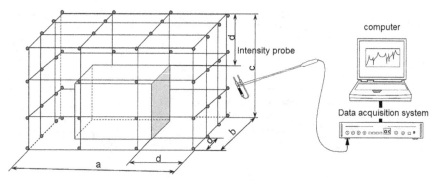

(a) Definition of the intensity measurement grid

(b) Measurement surfaces

Figure 6.43. Sound power measurements

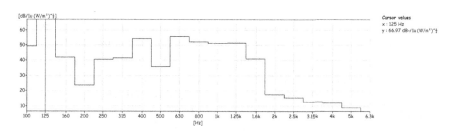

Figure 6.44. Third-octave band spectrum of the sound intensity radiated by the window surface without active control

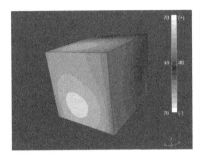

Figure 6.45. Sound intensity maps without active noise control

Figure 6.46. Detail of the main sound intensity paths

evaluate the effectiveness of the active noise control system in attenuating the sound power transmitted through the window at that frequency. Nevertheless, flanking sound transmission influenced the sound attenuation results. In the anechoic chamber, in fact, the sound transmitted through both the smart window and the neighboring walls of the sending room was measured. As shown in Fig. 6.46, although the opening in the upper side of the sending room was covered with sound absorbing material, a substantial fraction of the energy was transmitted via such a peripheral path, thus resulting in a degradation of the active noise control performance. For this reason, only the error sensor 2 positioned close to the window prototype was included in the control system architecture.

Figure 6.47 shows the spectrum of the sound intensity radiated by the smart panels with active control. The resulting sound intensity map is shown in Fig. 6.48 for the 125 Hz third-octave band. For the controlled

6.6. SOUND INTENSITY MEASUREMENTS

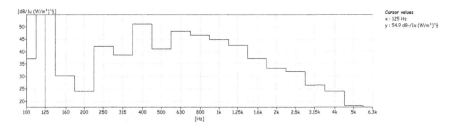

Figure 6.47. Third-octave band spectrum of the sound intensity radiated by the window surface with active control

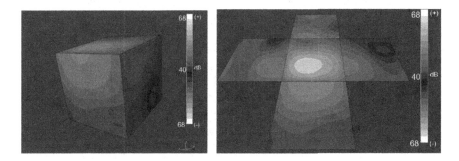

Figure 6.48. Sound intensity maps with active noise control

frequency range, the sound intensity in front of the window prototype is clearly reduced by the active noise control system resulting in a local reduction of 12.07 dB at the fundamental frequency. Moreover, as indicated in the local sound intensity distribution shown in Fig. 6.48, the control system is effective in reducing sound power at the window surface although the neighboring surfaces are poorly controlled. Sound transmission through the upper surface remains almost unchanged. In Fig. 6.49, the sound intensity maps of the radiating smart window surface are shown with and without control for the fundamental frequency of disturbance. Without control, the global mode (1,1) with the corresponding sound intensity is excited. With active control, the local intensity maximum is reduced to 55 dB, as shown in Fig. 6.47. Apparently, modal suppression plays a major role in modal restructuring.

Finally, a comparison between the sound power transmitted through all the measurement surfaces of the reference box with and without control is shown in Fig. 6.50. It is found that while controlling the noise transmission

238 CHAPTER 6. ACTIVE NOISE CONTROL EXPERIMENTS

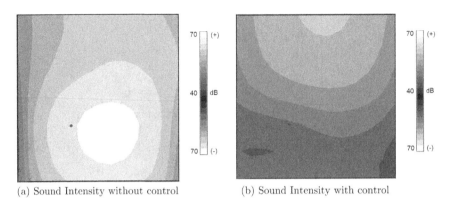

(a) Sound Intensity without control (b) Sound Intensity with control

Figure 6.49. Comparison of sound intensity third-octave band 125 Hz (dB) with and without control of the smart window

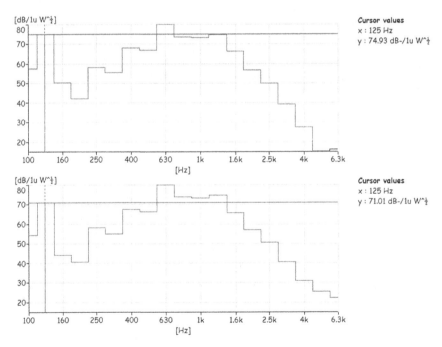

Figure 6.50. Comparison between the radiation soud power transmitted through all the measurement surfaces without (above) and with (below) control

through the window, the overall sound power radiated by the vibro-acoustic facility is also reduced in the same frequency range.

6.7 Passenger Comfort Estimation

Noise is one of the most important parameters affecting passenger comfort in aircraft. Active noise control concepts for NVH reduction therefore becomes most effective when the actual passenger perception of sound and vibration is taken into account. In this section, the noise impact on passengers is evaluated from the perspective of vibro-acoustic comfort. The effectiveness of the developed smart window prototype is examined not only in terms of noise and vibration attenuation but also by providing a quantitative estimate of comfort benefits at the passenger level. This is accomplished by relating the experimental noise attenuation results with an estimate of the subjective response perceived by passengers. To this aim, starting from the results available in the literature [182–184], a human response model to noise and vibration in aircraft cabin was specifically developed and implemented in an artificial neural network-based MATLAB® tool.

6.7.1 *Subjective noise perception*

Spectral time-fluctuations and amplitude modulations of a cabin vibro-acoustic environment can significantly affect the human body's perception of comfort. This correlation is not straightforward, as it may seem at first sight, because NVH perception is a subjective quality that depends not only on environmental conditions such as temperature, humidity, noise and vibration, but also on passenger conditions such as health, physiology and psychological attitudes.

It is well recognized that A-weighted measurements of sound pressure levels alone are insufficient in relation to sound quality and noise annoyance assessment. This has led to an increasing interest in subjective parameters in closer agreement with human sound perception.

As an acoustic stimulus can be described by objective parameters, like sound pressure level, frequency, duration, etc., the physical stimuli lead to different hearing sensations. Such physical magnitudes are correlated to psychophysical magnitudes such as loudness, pitch and subjective duration, usually referred to as hearing sensations. The relationship between the physical magnitude of the stimulus and the magnitude of the correlated hearing sensation is studied by psychoacoustics. Several psychoacoustic parameters characterize human subjective response. Loudness, sharpness, tonality, fluctuation strength and roughness, usually referred to as Zwicker's parameters, are the most recurrent [186]. Acoustic tonality, for instance, takes into account response peaks in narrow bands. It is therefore particularly indicated for cabin comfort evaluation in turboprop aircraft.

Modelling passenger comfort through mathematical tools is essential in order to target the human-centred design of a cabin interior. In this field, interesting results have been achieved in a number of research projects involving aircraft cabin comfort. In particular, IDEA-PACI (Identification of an Aircraft Passenger Comfort Index) succeeded in developing a specific mathematical tool capable of simulating and predicting passenger vibroacoustic comfort by a synthetic scalar index [13].

The psychoacoustic parameters were estimated from both objective and subjective perspectives: through elaboration of the physical output (acoustic response, hence objective) and elaboration of human answers (hence subjective). The resulting Virtual Passenger Model was experimentally validated in highly realistic aircraft cabin simulators by replicating cabin interior sound and vibration in different operating conditions. Figure 6.51 shows a sketch of a cabin simulator facility, typically used for passenger comfort investigations. More than 500 people were interviewed and a huge amount of data was generated. Their subjective comfort was statistically analyzed and elaborated by psychologists. The statistical distribution of subjective answers to different vibroacoustic environments was elaborated to generate a comprehensive model of cabin comfort perception. The input–output relation between physical environmental parameters, such as spectral and

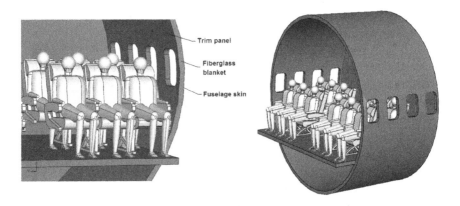

Figure 6.51. Sketch of a facility for passenger comfort tests

time-domain shapes of the simulated sound field, and the perceived comfort was modelled by an artificial neural network.

6.7.2 Virtual Passenger Model

In the framework of the IDEA-PACI project, the transfer function relating subjective acoustic stimuli to passenger perception of comfort was modelled by the so-called Virtual Passenger Model. A description of this tool can be found in [13]. An ANN-based model was developed by using JavaNNS® (Java Neural Network Simulator). The resulting Comfort Index (CI) in a turboprop aircraft cabin was described by a scalar index ranging in the interval [1, 7] and increasing with passenger discomfort.

Starting from the comfort data analysis and methods available in literature along with the corresponding results, the implementation of a MATLAB® code for simulating a virtual passenger response model to the aircraft vibro-acoustic environment was one of the goals of this book. This model was used to estimate the subjective interior comfort level by processing the active noise and vibration attenuation results obtained on the smart window test-bed.

The ANN architecture was based on a multi-layer perceptron with one hidden layer. The training was performed by using a back-propagation algorithm. Figure 6.52 shows the ANN training results describing the differences between the measured and the calculated comfort index values. The achieved results appear quite satisfactory: during the validation procedure, the gaps between the real values (measured) of the psycho-acoustic response and the simulated ones were widely inside an acceptable error range.

The potential benefits on passenger comfort from using the developed active noise control system were therefore evaluated. Microphone 2 was selected for a reliable description of the vibro-acoustic field perceived by passengers sitting next to the aircraft window. Figure 6.53 shows the frequency spectrum, plotted in third-octave bands, obtained by averaging the experimental vibro-acoustic data obtained for the narrow-band disturbance controlled by the MIMO control algorithm. Up to 18 dB of reduction was obtained at the selected frequency band. By comparing the experimental data with the active noise control simulations, described in Chapter 4, a perfect agreement can also be observed. These data were given as input to the developed ANN-based human response model. The comfort levels are reported in Fig. 6.54. In the subjective scale, the passenger comfort increases for decreasing CI values. This means that the CI can be better

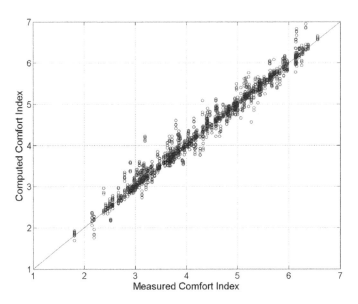

Figure 6.52. Measured vs. calculated Scalar Comfort Index in a turboprop aircraft cabin modelled by an artificial neural network

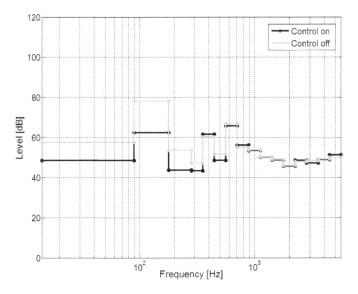

Figure 6.53. Average spectrum of the narrow-band signal measured at the error microphone 2 before and after the active noise control system operation using the MIMO control architecture

6.8. SUMMARY

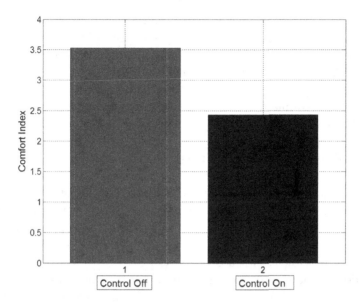

Figure 6.54. Vibro-acoustic Comfort Index results

described as a discomfort index: the lower the value, the better the noise and vibration from an annoyance point of view. Results show a successful improvement in the comfort level while turning the control system on. This demonstrates that such a system provides an alternative way to reduce cabin interior noise and consequently improve passenger comfort by selectively reducing the annoying components of sound and vibration at the passenger level.

6.8 Summary

In this chapter, real-time active noise control experimental strategies have been addressed. Specifically, tests have been performed on a benchmark structure to recommend practices for active noise control experiments in an acoustic test facility between a sending room and anechoic receiving room.

The control system has been based on commercially available piezoelectric stack actuators aimed at reducing sound radiation from a triple-panel partition. Practical implementations of SISO and MIMO adaptive filtered-X LMS control architectures have been realized to drive the vibration control sources. The secondary plant models have been identified off-line prior to the measurements. The controller has been implemented on a laboratory DSP control system by using the sum of squared pressures at discrete error sensing locations in the receiving room as error signal.

Different acoustic scenarios have been investigated by simulating the propeller induced interior noise of a turboprop aircraft in a reference test suite. The low-frequency modal properties of such a vibro-acoustic facility have been also evaluated. It has been tacitly assumed that a substantial attenuation in the sound transmitted through the experimental set-up led to an increase in the insertion loss of the smart structure. This assumption has been verified in the course of the experiments. The acoustic benefits have been successfully demonstrated with various artificially generated tonal and narrow-band disturbances exciting the experimental test-bed in stationary and transient conditions. A loudspeaker has been used as primary perturbation source with realistic acoustic levels.

A description of the noise attenuation results achieved by varying the number of control actuators and error sensors has been also provided. Up to four piezoelectric actuators have been controlled individually. Measurements obtained for different configurations have proved meaningful reductions in noise levels at the sensor positions. The capabilities of the adaptive structural acoustic controller to track changes in the noise disturbance have been tested against non-stationary acoustic perturbations. A spectogram of the active control performance has been provided for a frequency sweeping acoustic excitation.

Practical applications need a microphone spatial sampling in order to provide the controller with a proper estimate of the radiated acoustic energy. A canonical experimental procedure to measure the sound power radiated from the smart panels system has therefore been detailed. Besides the comparison between the output of the error sensors, the active reduction in the radiated sound power, averaged over the frequency band of interest, has been investigated by means of sound intensimetry.

Finally, although sound pressure level is the most used parameter in acoustic design and sound quality assessment, active noise control systems become really effective when the noise impact on passengers is evaluated from the point of view of comfort perception. For this reason, the performance of the smart panels system has been evaluated by taking the resulting passenger comfort parameters into account through a human response model predicting the perceived vibro-acoustic response benefits.

Appendix A
Benchmark Examples

A.1 Introduction

The use of piezoceramic patches as static and dynamic transducers is described in many technical papers. These devices are characterized by their quick reaction to fluctuations in voltages and strain due to the electromechanical coupling effect.

In this section, some numerical examples involving the modelling of structures with surface-bonded piezoelectric actuators and sensors are reported. First, the direct piezoelectric effect of a laminar piezoelectric sensor is modelled. The results are used in [187] for the numerical validation of a fast boundary element method for the analysis of damaged structures. Second, a static analysis is performed on a cantilever piezoelectric plate. Third, a test case of bimorph actuation is presented for a cantilever beam with two symmetrically attached piezoelectric patches. Both benchmark studies are carried out by using the piezoelectric finite element formulation implemented in ABAQUS®.

A.2 Laminar Piezoelectric Sensor

An isolated sensor with electroded surfaces is modelled first. The output voltages due to the application of prescribed strains at the base of the piezoelectric patch are numerically calculated by using ABAQUS®. Strains are imposed through prescribed displacements at the base of the sensor. The resulting voltage is computed for various values of thickness h_p of the sensor (10 mm × 10 mm × h_p) subjected to strains $S_{11} = S_{22} = 10^{-4}$ at the basis, as shown in Fig. A.1. The material constants are reported in Table A.1.

APPENDIX A. BENCHMARK EXAMPLES

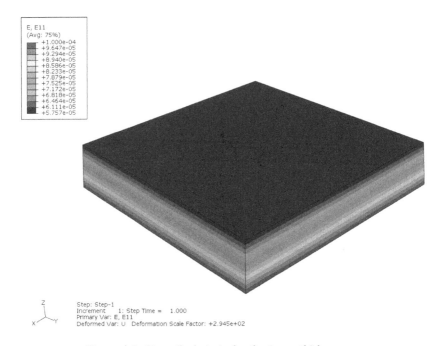

Figure A.1. Prescribed strain for the 1 mm thick sensor

Table A.1 Material constants for PZT-4

	C_{11}	C_{22}	C_{33}	C_{12}	C_{13}	C_{23}	C_{44}	C_{55}	C_{66}
GPa	139	139	115	77.80	74.30	74.30	25.60	25.60	30.60
	e_{31}	e_{32}	e_{33}	e_{24}	e_{15}				
C/m^2	−5.20	−5.20	15.08	12.72	12.72				
	ϵ_{11}	ϵ_{22}	ϵ_{33}						
nFa/m	13.06	13.06	11.51						

The output voltage for different sensor thickness is shown in Fig. A.2. Such results are compared with those computed by means of a simplified analytical model, presented by Lin and Yuan [185], and validated through experiments. Such a model predicts

$$V_{out} = \frac{d_{31} E_p h_p S_p}{\epsilon_{33}(1 - \nu_p)} \tag{A.1}$$

where $E_p = C_{11} = C_{22}$ and ν_p are the in-plane Young's modulus and Poisson's ratio of the piezoelectric material, h_p is the thickness of the piezoelectric patch, ϵ_{33} is the dielectric permittivity along the thickness direction

A.3. CANTILEVER PIEZOELECTRIC PLATE

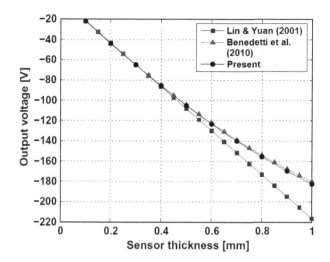

Figure A.2. Comparison between output voltages of numerical and analytical sensor for a benchmark configuration

and d_{31} is the piezoelectric constant relating S_{11} to E_3 or D_3 to T_{11} in the constitutive equations (5.3) and (5.4). Assuming that the sensor is sufficiently small and the strain is constant inside the sensor area, S_p is the value of the strain at the center of the piezoelectric patch.

The output voltages very well match the analytical 2D model for small values of h_p. For higher values of thickness, the effect of shear becomes more relevant. The FE model predictions are in perfect agreement with the numerical results presented in [187].

A.3 Cantilever Piezoelectric Plate

This section starts with the finite element investigation of a beam entirely made of piezoelectric material. Numerical results are compared with those reported by Mrad et al. in [188]. The reference structure has been validated against experiments. The piezoelectric plate is then bonded to an elastic beam. The local dynamic behaviour of the structure is investigated.

A.3.1 Static analysis

The length, width and thickness of the piezoelectric cantilever beam are 19.05 mm, 6.35 mm and 0.508 mm, respectively. Such a clamped beam is depicted in Fig. A.3.

APPENDIX A. BENCHMARK EXAMPLES

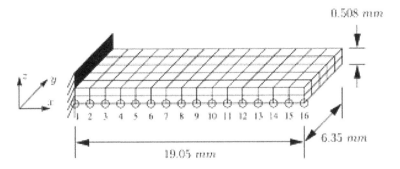

Figure A.3. Piezoceramic cantilever beam

Figure A.4. Numerical model of the piezoceramic cantilever beam

The structure is meshed with 150 ($15 \times 15 \times 2$) 8-node linear piezoelectric solid elements, shown in Fig. A.4.

The electromechanical properties of the piezoelectric cantilever beam, represented by the elastic, the piezoelectric and permittivity electric constants, are:

$$[c^E] = \begin{bmatrix} 12.1 & 7.54 & 7.52 & 0 & 0 & 0 \\ 7.54 & 12.1 & 7.52 & 0 & 0 & 0 \\ 7.52 & 7.52 & 11.1 & 0 & 0 & 0 \\ 0 & 0 & 0 & 2.26 & 0 & 0 \\ 0 & 0 & 0 & 0 & 2.26 & 0 \\ 0 & 0 & 0 & 0 & 0 & 2.11 \end{bmatrix} 10^{10} \text{ N/m}^2$$

$$[e] = \begin{bmatrix} 0 & 0 & 0 & 0 & 12.3 & 0 \\ 0 & 0 & 0 & 12.3 & 0 & 0 \\ -5.4 & -5.4 & 15.1 & 0 & 0 & 0 \end{bmatrix} \text{C/m}^2$$

$$[\epsilon^S] = \begin{bmatrix} 8.11 & 0 & 0 \\ 0 & 8.11 & 0 \\ 0 & 0 & 7.349 \end{bmatrix} 10^{-9} \text{ F/m}.$$

A.3. CANTILEVER PIEZOELECTRIC PLATE

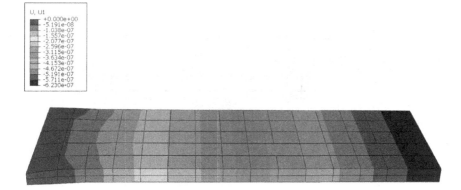

Figure A.5. Simulation results

Table A.2 Comparison between the displacements in the x-direction of the numerical and benchmark model

Node number	Node position (mm)	Displacements (m)		Error (%)
		FEM	Benchmark	
2	1.27	−1.81933 e-8	−1.71655 e-8	5.65
4	3.81	−1.20455 e-7	−1.16784 e-7	3.05
6	6.35	−2.08995 e-7	−2.06770 e-7	1.06
8	8.89	−2.92235 e-7	−2.92410 e-7	0.06
10	11.43	−3.74829 e-7	−3.77518 e-7	0.72
12	13.97	−4.57486 e-7	−4.62679 e-7	1.14
14	16.51	−5.40175 e-7	−5.47872 e-7	1.42
16	19.05	−6.22867 e-7	−6.33072 e-7	1.64

The displacements of the FEM nodes in the x-direction are computed by applying a voltage of 100 V. They are shown in Fig. A.5. In Table A.2, the achieved results are compared with those reported in [188], obtained by simulating the piezoelectric effect through the thermal analogy. A good agreement is achieved.

A.3.2 Dynamic analysis

The dynamic response of a piezoelectric actuator bonded to an aluminium cantilever beam is finally investigated. The structure is schematically described in Fig. A.6. A set of 408 8-node linear brick elements C3D8 is used to mesh the beam, whereas the piezoceramic is modelled with 80 8-node

250   APPENDIX A. BENCHMARK EXAMPLES

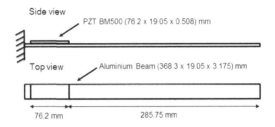

Figure A.6. Aluminum cantilever beam

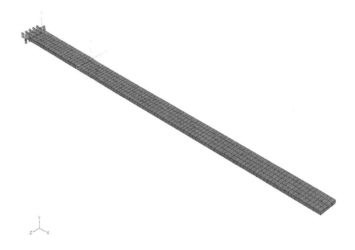

Figure A.7. FE model of the cantilever beam with a piezoelectric actuator

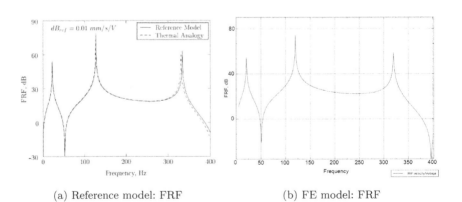

(a) Reference model: FRF   (b) FE model: FRF

Figure A.8. Comparison between the reference and the computed frequency response functions

A.3. CANTILEVER PIEZOELECTRIC PLATE

linear piezoelectric brick elements C3D8E, Fig. A.7. The elastic modulus and Poisson's ratio of the aluminium beam are 67 GPa and 0.33 respectively, while the density is 2730 kg/m^3. The piezoelectric properties are those used for the static analysis. A mass density of 7650 kg/m^3 is additionally considered. A constant actuation voltage of 0.2828 V is applied. The dynamic response of the structure is analyzed in the frequency range up to 400 Hz. The FE results are compared with the reference data reported in [188]. As shown in Fig. A.8, the frequency response function from the actuator voltage to the transverse velocity, computed at 211 mm from the clamped end, perfectly matches the reference curve.

Appendix B
Sound Power Simulations

In this section, the radiation efficiency of baffled rectangular plates is investigated. This benchmark study aims at validating the numerical procedure developed to predict the sound power radiation of vibrating structures. The FEM-based model of sound radiation through elemental radiators is compared with the results obtained by a standard BEM commercial code, i.e. Sysnoise®, used as a validation tool.

B.1 Radiation Efficiency

The radiation efficiency of the structural modes of a simply supported rectangular plate ($L_x = 0.2$ m, $L_y = 0.32$ m) mounted in an infinite baffle is determined by the sound power radiated by the computed velocity profile normalized by the sound power radiated by a rigid piston radiating with the same average velocity. The sound power radiation is predicted by using the standard boundary element formulation implemented in Sysnoise®. The structural modes are numerically calculated in ABAQUS®. The acoustic field is divided into two completely separate regions by a rigid baffle, simulating the noise transmission between a reverberant chamber and an anechoic chamber. A perfect coincidence between the structural and acoustic meshes is assumed, Fig. B.1. Transparent conditions are imposed to the elements in the plane of the baffle, in order to make them completely transparent to the emitted acoustic waves, Fig. B.2.

As shown in Fig. B.3, the sound power radiated from the structure is computed by integrating the far-field intensity over a hemispherical surface centred on the panel. Figure B.4 shows the sound power radiated into the upper half space by the mode (1,1) at the coincidence frequency.

APPENDIX B. SOUND POWER SIMULATIONS

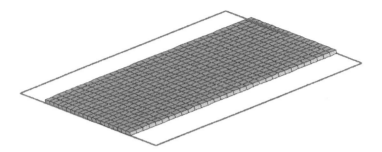

Figure B.1. Coupled structure and acoustic mesh

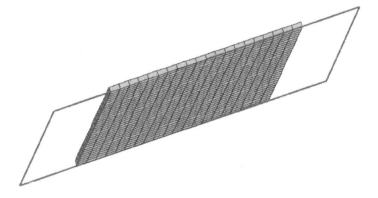

Figure B.2. Acoustic box mesh with "Transparency Plane" over the acoustic baffle

The radiation efficiency of the (1,1), (2,1) and (3,1) modes of the considered panel is shown in Fig. B.5. The approximate expressions, given by Eqs. (2.84)–(2.86), are also plotted. It can be seen that in the low frequency region, when the size of the panel is low compared with the acoustic wavelength, the (1,1) mode and the (3,1) mode radiate considerably more efficiently than the (2,1) mode. These results are consistent whith those published in [81]. The important difference between the (1,1) and (3,1) modes and the (2,1) mode is that when the panel is vibrating in the (2,1) mode at low frequencies, most of the air in front of the panel is transferred from one side of the panel to the other, generating little sound. In contrast, due to the net (space-average) volumetric velocity, the (odd,odd) modes displace fluid into the medium while vibrating. This fluid motion generates sound.

B.2. RADIATION EFFICIENCY OF THE WINDOW TEST-BED

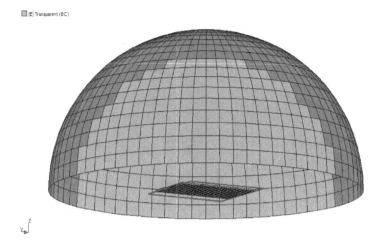

Figure B.3. Field point mesh for the evaluation of the sound power radiation

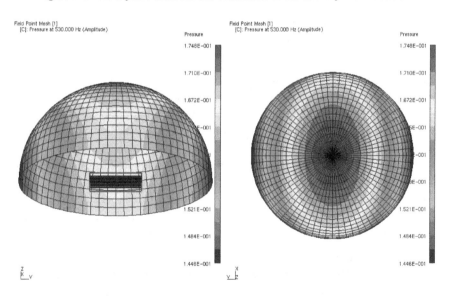

Figure B.4. Sound pressure radiated by the mode (1,1) of the simply supported panel

B.2 Radiation Efficiency of the Window Test-Bed

In what follows, the radiation efficiency of the low-frequency modes of the plexiglass plate of the window test-bed is presented. Since the vibro-acoustic

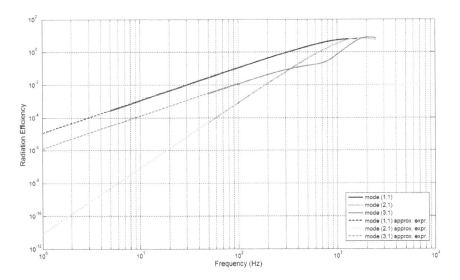

Figure B.5. Radiation efficiency of simply supported panel modes

properties of the window assembly are significantly influenced by the plexiglass panel, it is assumed that the sound power radiated at low frequencies is due to the vibration of the radiating panel, i.e. the outer plexiglass panel. This assumption was experimentally validated in Chapter 5. The baffle provides the only means of wave transmission from one half of the acoustic field to the other.

Figure B.6 shows the radiation efficiency of the structural modes, or self-radiating terms of the radiation matrix, of the radiating panel. The geometry used in this simulation corresponds to the experimental set-up. The abscissa of the plot is the acoustic wavenumber k_a normalized by the individual structural wave numbers $k_s = \sqrt{(\frac{p\pi}{L_x})^2 + (\frac{q\pi}{L_y})^2}$ to emphasize the critical frequencies, where the structural and acoustic wave numbers are equivalent. Sufficiently below the critical frequency, where $k_a/k_s \ll 1$, each radiation efficiency curve is nearly a straight line on a log-log scale. Just above the critical frequency, where $k_a/k_s > 1$, each radiation efficiency curve displays a peak, and the height increases with the mode numbers. Well above the critical frequency, where $k_a/k_s \gg 1$, the curves asymptotically approach unity. Comparison of the results achieved with the developed technique and Sysnoise® shows a similar behaviour.

B.2. RADIATION EFFICIENCY OF THE WINDOW TEST-BED

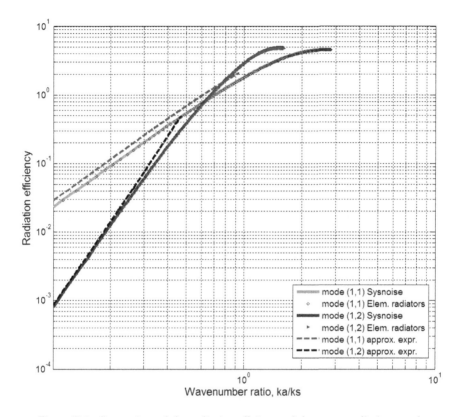

Figure B.6. Comparison of the radiation efficiency of the smart radiating panel

It is important to note that the radiation coupling between modes, represented by the off-diagonal terms in the radiation matrix, which are not shown in Fig. B.6, cannot be neglected when representing the total radiated acoustic power. Neglecting coupling terms greatly underestimates the radiated sound power.

Appendix C

Preliminary Experimental Studies

Preliminary experimental studies to validate the piezo stack-based actuator concept and identify practical limitations are detailed herein. To this aim, a rectangular plate was manufactured and tested. Two PICA stacks, of dimensions 11 × 16 mm and specifically designed for high-duty-cycle applications, were used [178]. The experimental set-up is illustrated in Fig. C.1. Tests were carried out by the authors at the Centre for Acoustics and Vibration (CAV) of Penn State University and were funded by the Italian Aerospace Research Centre (CIRA).

An experimental modal analysis has been carried out by using two different techniques: a MISO approach and a Scanning Laser Vibrometer. The

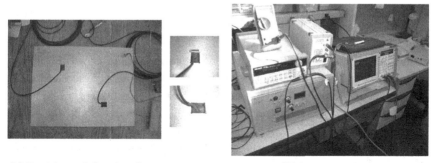

(a) Locations of the piezoelectric actuators   (b) Instrumentation

Figure C.1. Experimental set-up

260   APPENDIX C. PRELIMINARY EXPERIMENTAL STUDIES

Figure C.2. Scanning Laser Vibrometer

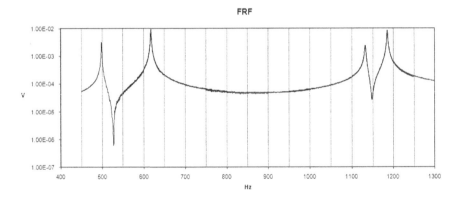

Figure C.3. Experimental frequency response function

former allows the determination of the structural resonances, within the frequency range of interest, from the peaks measured by a single accelerometer placed at the plate corner. The sensor is shown in Fig. C.1(a). The latter guarantees a more accurate measure of both the resonances and the structural mode shapes. The experimental investigation has been carried out in the range 450–1300 Hz. A chirp signal (swept sine) has been used to drive the piezoelectric actuators with the same intensity at all frequencies.

The structural modal parameters have been extracted by processing the frequency response functions from the actuator voltage to the normal sensor

Table C.1 Numerical vs. experimental results

	Numerical results		Experimental results	
Mode	C3D8R elements (Hz)	C3D20 elements (Hz)	Single Accelerometer (Hz)	Laser Vibrometer (Hz)
I	476.44	494.18	498.75	498.8
II	580.58	603.01	617.5	621.3
III	1053.90	1094.30	1132	1135
IV	1124.50	1166.60	1185.25	1190

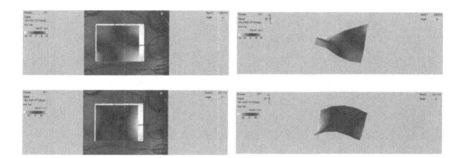

Figure C.4. Experimental mode shapes

displacements. A MDOF (multi-degree of freedom) curve fitting method has been applied. The distribution of the mass-loading-free vibration of the structure has been measured without physical contact with the test object. The structural mode shapes have been experimentally measured in 49 points of acquisition through a Polytec PSV-400 Scanning Laser Vibrometer, shown in Fig. C.2. The measurement principle is based on the Doppler Effect.

In Fig. C.3, the average of the measured FRFs is shown. The structural resonances are associated with the main peaks of the curve. As illustrated in Table C.1, the experimental results match the numerical simulations very well. The two-dimensional and three-dimensional representations of the experimental mode shapes are finally illustrated in Fig. C.4.

Bibliography

[1] R. von der Bank, S. Donnerhack, A. Rae, F. Poutriquet, A. Lundbladh, and A. Schweinberger. Breakthrough core-engine technologies for emission reduction. In *Proceedings of the 3AF/CEAS Greener Aviation Conference*, 2014.

[2] E. Kors. Roadmap brick for aircraft noise reduction. In *Proceedings of the 3AF/CEAS Greener Aviation Conference*, 2014.

[3] A. Krien. The future by airbus. In *Proceedings of the Green Aviation Symposium*, 2011.

[4] T. Revel. Integrated lightweight cmc active/adaptive nozzle. In *Proceedings of the 3AF/CEAS Greener Aviation Conference*, 2014.

[5] C. Seror-Goguet. Airframe noise overview. In *Proceedings of the 3AF/CEAS Greener Aviation Conference*, 2014.

[6] M. Smith. OPENAIR integrated propulsion system design summary. In *Proceedings of the 3AF/CEAS Greener Aviation Conference*, 2014.

[7] M. Bauer. Novel noise reduction technologies. In *Proceedings of the 3AF/CEAS Greener Aviation Conference*, 2014.

[8] A. Mann, F. Perot, M.S. Kim, and D. Casalino. Characterization of acoustic liners absorption using a lattice-Boltzmann method. In *Proceedings of the 19th AIAA/CEAS Aeroacoustics Conference*, 2013.

[9] M. Barbarino, I. Dimino, A. Carozza, C. Nae, C. Stoica, V. Pricop, S.H. Peng, P. Eliasson, O. Grundestam, L. Tysell, L. Davidson, L.E. Eriksson, H.D. Yao, S. Ben Khelil, F. Moens, T. Le Garrec, D.C. Mincu, F. Simon, E. Manoha, J.L. Godard, and M.A. Averardo. Airframe noise reduction technologies applied to high-lift devices of future green regional aircraft. In *Proceedings of the 3AF/CEAS Greener Aviation Conference*, 2014.

[10] I. Dimino, D. Flauto, G. Diodati, A. Concilio, and R. Pecora. Actuation system design for a morphing wing trailing edge. *Recent Patents on Mechanical Engineering*, 7(3):1–11, 2014.

[11] S. Kota, R. Osborn, G. Ervin, and D. Maric. Mission adaptive compliant wing design, fabrication and flight test. In *NATO Research and Technology Organization, RTO-MP-AVT-168*, pp. 1–19, 2009.

[12] ISO 2631. Mechanical vibration and shock — Evaluation of human exposure to whole-body vibration. 2004.

[13] http://www.idea-paci.org. [Accessed on March 16, 2014].

[14] J.F. Wilby. Aircraft interior noise. *Journal of Sound and Vibration*, 190(3):545–564, 1996.

[15] L. Cremer, M. Heckl, and E. Ungar. *Structure-Borne Sound*. Berlin: Springer-Verlag, 1988.

[16] J.F. Piet, N. Molin, and C. Sandu. Aircraft landing gear provided with at least one noise reducing means. *Patent N. 8256702*, United States Patent, 2012.

[17] M. Murayama, Y. Yokokawa, and K. Yamamoto. Computational study on noise generation from a two-wheel main landing gear. In *Proceedings of ICAS 2012, 28th International Congress of the Aeronautical Science*, 2012.

[18] J. Morkholt, J. Hald, P. Hardy, D. Trentin, M. Bach-Andersen, and G. Keith. Measurement of absorption coefficient, surface admittance, radiated intensity and absorbed intensity on the panels of a vehicle cabin using a dual layer array with integrated position measurement. *SAE International Journal of Passenger Cars–Mechanical Systems*, 2:1449–1457, October 2009.

[19] A. Brancati. *Boundary element method for fast solution of acoustic problems: Active and passive noise control*. Ph.D. Thesis, Imperial College London, 2010.

[20] B.N. Tran, G.P. Mathur, and P.G. Bremner. Modal energy analysis of aircraft sidewall response to acoustic fields. In *Proceedings of the Internoise Conference*, 1995.

[21] E. Waterman, D. Kaptein, and S. Sarin. Fokker's activities in cabin noise control for propeller aircraft. In *Proceedings of the SAE Business Aircraft Meeting and Exposition*, 1983.

[22] K.A. Cunefare and G.H. Koopmann. Global optimum active noise control: surface and far-field effects. *Journal of the Acoustical Society of America*, 90:365–373, 1991.

[23] A. Concilio, L. Lecce, and A. Ovallesco. Position and number optimization of actuators and sensors in active noise control system by genetic algorithms. In *Proceedings of the AIAA/CEAS Aeroacoustics Conference*, 1995.

[24] J. Lefevre and U. Gabbert. Finite element modelling of vibro-acoustic systems for active noise reduction. *Technische Mechanik*, 25:241–247, 2005.

[25] H. Breitbach, C. Gerner, and D. Sachau. Device and method for reducing sound of a noise source in narrow frequency ranges. *Patent N. US 7,885,415 B2*, United States Patent, 2011.

[26] D. Guyomar, T. Richard, and C. Richard. Sound wave transmission reduction through a plate using a piezoelectric synchronized switch damping technique. *Journal of Intelligent Material Systems and Structures*, 19(7):791–803, 2008.

[27] W. Larbi, J.F. Deu, M. Ciminello, and R. Ohayon. Structural-acoustic vibration reduction using switched shunt piezoelectric patches: A finite element analysis. *Journal of Vibration and Acoustics*, 132(5):051006, 2010.

[28] E. Lefeuvre, A. Badel, L. Petit, C. Richard, and D. Guyomar. Semi-passive piezoelectric structural damping by synchronized switching on voltage source. *Journal of Intelligent Material Systems and Structures*, 17:653–660, 2006.

[29] G.A. Lesieutre. Vibration damping and control using shunted piezoelectric materials. *The Shock and Vibration Digest*, 30(3):187–195, 2008.

[30] O. Thomas, J.F. Deu, and J. Ducarne. Vibrations of an elastic structure with shunted piezoelectric patches: Efficient finite element formulation and electromechanical coupling coefficients. *International Journal for Numerical Methods in Engineering*, 80(2):235–268, 2009.

[31] P. Gardonio. Review of active techniques for aerospace vibro-acoustic control. *Journal of Aircraft*, 39(2):206–214, 2002.

[32] W.T. Baumann, F. Ho, and H.H. Robertshaw. Active structural acoustic control of broadband disturbances. *Journal of the Acoustical Society of America*, 92(4):1998–2005, 1992.

[33] R.L. Clark and C.R. Fuller. Optimal placement of piezoelectric actuators and polyvinylidene fluoride (PVDF) error sensors in active structural acoustic control approaches. *Journal of the Acoustical Society of America*, 92(3):1489–1498, 1992.

[34] R.L. Clark and C.R. Fuller. Control of sound radiation with adaptive structures. *Journal of Intelligent Material Systems and Structures*, 2(3):431–452, 1991.

[35] D.G. MacMartin. Collocated structural control for reduction of aircraft cabin noise. *Journal of Sound and Vibration*, 190(1):105–119, 1996.

[36] L. Meirovitch and S. Thangjitham. Active control of sound radiation pressure. *Journal of Vibration, Acoustics, Stress and Reliability in Design*, 112(2):237–244, 1990.

[37] H.T. Banks, R.J. Silcox, and R.C. Smith. Approximation methods for control of structural acoustics models with piezoceramic actuators. *Journal of Intelligent Materials Systems and Structures*, 4:98–116, 1993.

[38] Deyu Li. *Vibroacoustic behaviour and noise control studies of advanced composite structures*. Ph.D. Thesis, University of Pittsburgh, 2003.

[39] K. Wolf, S. Han, T. Doll, L. Kurtze, T. Bein, and H. Hanselka. Performance assessment of active noise reduction systems. In *Proceedings of the 9th Biennial ASME Conference on Engineering Systems Design and Analysis (ESDA)*, 2008.

[40] R.J. Silcox, C.R. Fuller, and H.C. Lester. Mechanisms of active control in cylindrical fuselage structures. *AIAA Journal*, 28(8):1397–1404, 1990.

[41] D.J. Rossetti, M.A. Norris, S.C. Southward, and J.Q. Sun. A comparison of speakers and structural-based actuators for aircraft cabin noise control. In *Proceedings of the Second Conference on Recent Advances in Active Control of Sound and Vibration*, 1993.

[42] A. Arnhym. *Comfortization of Aircraft*. New York: Pitman Publishing Corporation, 1944.

[43] R.J. Elliott, P.A. Nelson, I.M. Stothers, and C.C. Boucher. In-flight experiments on the active control of propeller-induced cabin noise. *Journal of Sound and Vibration*, 140(2):219–238, 1990.

[44] C.R. Fuller and R.J. Silcox. Acoustics 1991: Active structural acoustic control. *Journal of the Acoustical Society of America*, 91(1):519, 1992.

[45] E. Monaco, L. Lecce, C. Natale, S. Pirozzi, and C. May. Active noise control in turbofan aircrafts: Theory and experiments. In *Acoustics 08*, pages 4629–4634, 2008.

[46] S. Johansson. *Active control of propeller-induced noise in aircraft*. Ph.D. Thesis, Blekinge Institute of Technology, Ronneby, Sweden, 2000.

[47] W. Halvorsen and U. Emborg. Interior noise control of the Saab 340 aircraft. In *Proceedings of the SAE General Aviation Aircraft Meeting and Exposition*, 1989.

[48] A. Grewal, D.G. Zimcik, and B. Leigh. Feedforward piezoelectric structural control: An application to aircraft cabin noise reduction. *Journal of Aircraft*, 38(1):164–173, 2001.

[49] Ultra Electronics Ltd. http://www.ultra-electronics.com [Accessed on December 1, 2014].

[50] P. Konstanzer, M. Grunewald, P. Janker, and S. Storm. Aircraft interior noise reduction through a piezo tunable vibration absorber system. In *Proceedings of the International Congress of the Aeronautical Sciences*, 2006.

[51] O. Callsen, O. Von Estorff, W. Gleine, and S. Lippert. Investigation of the vibro-acoustic behaviour of aircraft cabin panels using FEM and BEM. In *Proceedings of the LMS User Conference*, 2007.

[52] L. Guillaumie, J.M. David, and O. Voisin Grall. Internal fuselage noise: Comparison of computations and tests on a falcon airframe. *Journal of Aerospace Science and Technology*, 4(8):545–554, 2000.

[53] R.E. Hayden, B.S. Murray, and M.A. Theobald. A study of interior noise levels, noise sources and transmission paths in light aircraft. *NASA CR 172152*, 1985.

[54] R.A. Prydz and F.J. Balena. Window acoustic study for advanced turboprop aircraft. *NASA CR 172391*, 1984.

[55] F. Unruh James and D. Till Paul. General aviation interior noise: Part 3, noise control measure evaluation. *NASA/CR-2000-211667*, 2002.

[56] V. Mellert, I. Baumann, N. Freese, and R. Weber. Impact of sound and vibration on health, travel comfort and performance of flight attendants and pilots. *Aerospace Science and Technology*, 12(1):18–25, 2008.

[57] J.K. Henry and R.L. Clark. Active control of sound transmission through a curved panel into a cylindrical enclosure. *Journal of Sound and Vibration*, 249(2):325–349, 2002.

[58] O.C. Zienkiewicz, R.L. Taylor, J.Z. Zhu, and P. Nithiarasu. *The Finite Element Method*. Oxford, UK: Butterworth-Heinemann, 2005.

[59] R. Ohayon and C. Soize. *Structural Acoustics and Vibration*. London, UK: Academic Press, 1998.

[60] M.H. Aliabadi. *The Boundary Element Method*. London, UK: John Wiley and Sons, 2002.

[61] V. Mallardo and M.H. Aliabadi. An adaptive fast multipole approach to 2D wave propagation. *Computer Modeling in Engineering and Sciences*, 87(2):77–96, 2012.

[62] N. Atalla and R.J. Bernhard. Review of numerical solutions for low-frequency structural-acoustic problems. *Applied Acoustics*, 43(2):271–294, 1994.

[63] D.J. Nefske, J.A. Wolf, and L.J. Howell. Structural-acoustic finite element analysis of the automobile passenger compartment: A review of current practice. *Journal of Sound and Vibration*, 80(2):247–266, 1982.

[64] J. Lefevre and U. Gabbert. Finite element simulation of smart lightweight structures for active vibration and interior acoustic control. *Technische Mechanik*, 23(1):59–69, 2003.

[65] J.F. Deu, W. Larbi, and R. Ohayon. Piezoelectric structural acoustic problems: Symmetric variational formulations and finite element results. *Computer Methods in Applied Mechanics and Engineering*, 197(19–20):1715–1724, 2008.

[66] K. Vorgote, B. Van Genechten, D. Vandepitte, and W. Desmet. On the analysis of vibro-acoustic systems in the mid-frequency range using a hybrid deterministic-statistical approach. *Computers and Structures*, 89(11–12):868–877, 2011.

[67] Lueg, P. Process of silencing sound oscillations. Patent No. 2043416. US Patent 1936.

[68] H.F. Olson and E.G. May. Electronic sound absorber. *Journal of the Acoustical Society of America*, 25:1130-1136, 1953.

[69] F. Fahy and P. Gardonio. *Sound and Structural Vibration: Radiation, Transmission and Response*. London, UK: Academic Press, 2007.

[70] ASTM Standard. *ASTM E90-04 Standard Test Method for Laboratory Measurement of Airborne Sound Transmission Loss of Building Partitions and Elements*. American Society for Testing and Materials, 2004.

[71] ASTM Standard. *ASTM E2249-02 (08) Standard Test Method for Laboratory Measurement of Airborne Sound Transmission Loss of Building Partitions and Elements using Sound Intensity*. American Society for Testing and Materials, 2008.

[72] D.J. Ewins. *Modal Testing: Theory and Practice*. London, UK: John Wiley and Sons, 1984.

[73] G. Gloth and P. Lubrina. Ground vibration test of the Airbus A380-800. In *Proceedings of the International Forum on Aeroelasticity and Structural Dynamics (IFASD)*, 2005.

[74] D. Goge, M. Boswald, U. Fullekrug, and P. Lubrina. Ground vibration testing of large aircraft — state-of-the-art and future perspectives. In *Proceedings of the XXV International Modal Analysis Conference (IMAC)*, 2007.

[75] C. Beatrix. Experimental determination of the vibratory characteristics of structures. *ONERA Technical Report n. 212*, 1973.

[76] N. Maia, J. Silva, J. He, N. Lieven, R. Lin, G. Skingle, W. To, and A. Urgueira. *Theoretical and Experimental Modal Analysis*. Hertfordshire, UK: Research Studies Press Ltd., 1998.

[77] P. Verboven. *Frequency-domain system identification for modal analysis*. Ph.D. Thesis, Vrije University Brussel, 2002.

[78] LMS International News. It's a "go" for the unmanned space vehicle at CIRA. http://epub01.publitas.nl/56/7/magazine.php.page=18 language=EN, 2010 [Accessed on December 1, 2014].

[79] D.A. Bies and C.H. Hansen. *Engineering Noise Control: Theory and Practice*. Abingdon, UK: Taylor and Francis Group, 2009.

[80] P.A Nelson and S.J. Elliott. *Active Control of Sound*. London, UK: Academic Press, 1993.

[81] F. Fahy. *Sound and Structural Vibration*. London, UK: Academic Press, 1985.

[82] C.E. Wallace. Radiation resistance of a rectangular panel. *Journal of the Acoustical Society of America*, 51(3): Part 2, 946-952, 1972.

[83] J.P. Arenas. Matrix method for estimating the sound power radiated from a vibrating plate for noise control engineering applications. *Latin American Applied Research*, 4:345–352, 2009.

[84] J. Lee and J. Ih. Significance of resonant sound transmission in finite single partitions. *Journal of Sound and Vibration*, 277:881–893, 2004.

[85] A. Putra and D.J. Thompson. Sound radiation from rectangular baffled and unbaffled plates. *Applied Acoustics*, 71(12):1113–1125, 2010.

[86] J.P. Arenas. Numerical computation of the sound radiation from a planar baffled vibrating surface. *Journal of Computational Acoustics*, 16:321–341, 2008.

[87] P. Vitiello, P.A. Nelson, and M. Petyt. Numerical studies of the active control of sound transmission through double partitions. *ISVR Technical Report No. 183*, 1989.

[88] S.J. Elliott and M.E. Johnson. Radiation modes and the active control of sound power. *Journal of the Acoustical Society of America*, 94(4):2194–2204, 1993.

[89] S.D. Snyder and N. Tanaka. Calculating total acoustic power output using modal radiation efficiencies. *Journal of the Acoustical Society of America*, 97(3):1702–1709, 1995.

[90] P.M. Morse and U. Ingard. *Theoretical Acoustics*. New York, NY: McGraw-Hill, 1968.

[91] A.D. Pierce. *Acoustics: An Introduction to its Physical Principles and Applications*. New York, NY: Acoustical Society of America, 1989.

[92] G.V. Borgiotti. The power radiated by a vibrating body in an acoustic fluid and its determination from boundary measurements. *Journal of the Acoustical Society of America*, 88:1884–1893, 1990.

[93] S. Elliott. *Signal Processing for Active Control*. London, UK: Academic Press, 2000.

[94] F. Fahy and J. Walker. *Advanced Applications in Acoustic, Noise and Vibration*. London, UK: London Spon Press, 2004.

[95] R.Y. Vinokur. Transmission loss of triple partitions at low frequencies. *Applied Acoustics*, 29:15–24, 1990.

[96] R.S. Puri, D. Morrey, and J.F. Durodola. A comparison of structural-acoustic coupled reduced order models (ROMS): Modal coupling and implicit moment matching via arnoldi. In *Proceedings of the XXV International Modal Analysis Conference (IMAC)*, 2007.

[97] S. Marburg. Developments in structural-acoustic optimization for passive noise control. *Archives of the Computational Methods in Engineering*, 9(4):291–370, 2002.

[98] R. Vacaitis. Study of noise transmission through double wall aircraft windows. *NASA CR 172182*, 1983.

[99] W. Desmet and P. Sas. Vibro-acoustic analysis of the low frequency insertion loss of finite double-panel partitions. In *Proceedings of the International Congress MV2*, 1995.

[100] J.P. Carneal and C.R. Fuller. An analytical and experimental investigation of active structural acoustic control of noise transmission through double panel systems. *Journal of Sound and Vibration*, 272(3–5):749–771, 2004.

[101] O.E. Kaiser, S.J. Pietrzko, and M. Morari. Feedback control of sound transmission through a double glazed window. *Journal of Sound and Vibration*, 263:775–795, 2003.

[102] ABAQUS, Analysis (2009). User's Manual Version 6.9. *Dassault Systemes Simulia Corp. Providence*, 2009.

[103] A. Peiffer, S. Bruhl, C. Moeser, and S. Tewes. Zeitoptimierte berechnung des schalldurchgangs durch wandstrukturen mittels FEM. Fortschritte der Akustik, 31:135, 2005.

[104] A.J. Van Engelen. Sound radiation of a baffled plate: Theoretical and numerical approach. In *Research Report, Eindhoven University of Technology Department of Mechanical Engineering Dynamics and Control Group*, 2009.

[105] B. Tiseo, I. Dimino, A. Gianvito, and A. Concilio. A broadband acoustic feedback control system on confined area on board of aircraft. In *Proceedings of the 13th AIAA/CEAS Aeroacoustics Conference*, 2007.

[106] A. Preumont. *Vibration Control of Active Structures*. Dordrcht, Netherlands: Kluwer Academic Publishers, 2002.

[107] M.A. Galland and B. Mazeaud. Decentralized feedback control for active absorption in flow ducts. In *Proceedings of ACTIVE 2006: The 2006 International Symposium on Active Control of Sound and Vibration*, 2006.

[108] I.Z. Mat Darus and M.O. Tokhi. Soft computing-based active vibration control of a flexible structure. *Engineering Applications of Artificial Intelligence*, 18(1):93–114, 2005.

[109] O. Pabst, T. Kletschkowski, and D. Sachau. Active noise control in light jet aircraft. *Journal of The Acoustical Society of America*, 123(5):3875, 2008.

[110] T. Kletschkowski, D. Sachau, S. Bohme, and H. Breitbach. Optimized active noise control of semiclosed aircraft interiors. *International Journal of Aeroacoustics*, 6:61–71, 2007.

[111] K. Kochan, T. Kletschkowski, D. Sachau, and H. Breitbach. Active noise control in a semi-closed aircraft cabin. *Noise & Vibration Worldwide*, 41:8–20, 2010.

[112] K. Kochan, D. Sachau, and H. Breitbach. Robust active noise control in the loadmaster area of a military transport aircraft. *Journal of the Acoustical Society of America*, 129(5):3011, 2011.

[113] J.P. Maillard and C.R. Fuller. Active control of sound radiation from cylinders with piezoelectric actuators and structural-acoustic sensing. *Journal of Sound and Vibration*, 222:363–388, 1999.

[114] J.S. Vipperman, R.A. Burdisso, and C.R. Fuller. Active control of broadband structural vibration using the LMS adaptive algorithm. *Journal of Sound and Vibration*, 166:283–299, 1993.

[115] M.O. Tokhi and S. Veres. *Active Sound and Vibration Control — Theory and Applications*. London, UK: The Institution of Electrical Engineers, 2002.

[116] M. De Diego and A. Gonzalez. Performance evaluation of multichannel adaptive algorithms for local active noise control. *Journal of Sound and Vibration*, 244(4):615–634, 2001.

[117] J. Minkoff. The operation of multichannel feedforward adaptive systems. *IEEE Transactions on Signal Processing*, 45:2993–3005, 1997.

[118] F. Grosveld, R. Navaneethan, and J. Roskam. Noise reduction characteristics of general aviation type dual pane windows. In *Proceedings of the AIAA Aircraft Systems and Technology Meeting*, 1980.

[119] M. Rassaian, J.C. Lee, J. Montgomery, P.S. Guard, M.A. Stadum, P.E. Mikulencak, and R. Brigman. Optimal aircraft window shape for noise control. Patent No. 7438263 B2, US Patent 2008.

[120] K. Hills, S. Wong, M. Pare, W. Miller, N. Waidner, and R. Mumm. Integrated window belt system for aircraft cabins. In Patent No. 7118069 B2, US Patent 2006.

[121] K. Kochan and D. Sachau. Active noise reduction. In *dSPACE NEWS 2007/2*, 2007.

[122] J. Pyper. Laminated glass provides noise barrier benefits in automobile and architectural applications. *Journal of Sound and Vibration*, 35:8, 10–16, 2001.

[123] G.E. Freeman and R.A. Esposito. Glazing for vehicle cabin sound reduction. In *Proceedings of Internoise 2002*, 2002.

[124] J.G. Lilly. Recent advances in acoustical glazing. *Journal of Sound and Vibration*, 38:8–13, 2004.

[125] J. Lu. Passenger vehicle interior noise reduction by laminated side glass. In *Proceedings of Internoise 2002*, 2002.

[126] R. Buehrle, G. Gibbs, J. Klos, and A. Sherilyn. Noise transmission characteristics of damped plexiglas windows. In *Proceedings of the 8th AIAA/CEAS Aeroacoustics Conference, June 16–18*, 2002.

[127] R. Buehrle, G. Gibbs, and J. Klos. Damped windows for aircraft interior noise control. In *Proceedings of NOISE-CON 2004, July*, 2004.

[128] R. Buehrle, G. Gibbs, J. Klos, and M. Mazur. Modeling and validation of damped plexiglas windows for noise control. In *Proceedings of the AIAA/ASME/ASCE/AHS/ASC Structures, Structural Dynamics, and Materials Conference, Norfolk, Virginia, AIAA 2003-1870*, 2003.

[129] G.P. Gibbs, R.D. Buehrle, J. Klos, and S.A. Brown. Noise transmission characteristics of damped plexiglas windows. In *Proceedings of the 8th AIAA/CEAS Aeroacoustics Conference, June 16–18*, 2002.
[130] J.D. Quirt. Sound transmission through windows: I. Single and double glazing. *Journal of the Acoustical Society of America*, 72(3):834–844, 1982.
[131] J.D. Quirt. Sound transmission through windows: II. Double and triple glazing. *Journal of the Acoustical Society of America*, 74(2):534–542, 1983.
[132] A. Brekke. Calculation methods for the transmission loss of single, double and triple partitions. *Applied Acoustics*, 14:225–240, 1981.
[133] A.J.B. Tadeu and D.M.R. Mateus. Sound transmission through single, double and triple glazing. experimental evaluation. *Applied Acoustics*, 62:307–325, 2001.
[134] F.X. Xin, T.J. Lu, and C.Q. Chen. Sound transmission through simply supported finite double-panel partitions with enclosed air cavity. *Journal of Vibration and Acoustics*, 132(1):011008, 2010.
[135] F.X. Xin and T.J. Lu. Analytical modeling of sound transmission across finite aeroelastic panels in convected fluids. *Journal of The Acoustical Society of America*, 128(3):1097–1107, 2010.
[136] T. Wang, S. Li, S. Rajaram, and S.R. Nutt. Predicting the sound transmission loss of sandwich panels by statistical energy analysis approach. *Journal of Vibration and Acoustics*, 132, 2010.
[137] R. Vaicaitis. Study of noise transmission through double wall aircraft windows. *NASA CR 172182*, 1983.
[138] F. Grosveld. Acoustic transmissibility of advanced turboprop aircraft windows. *Journal of Aircraft*, 25(10):942–947, 1988.
[139] J.P. Carneal and C.R. Fuller. Active structural acoustic control of noise transmission through double panel systems. *AIAA Journal*, 33(4):618–623, 1995.
[140] A. Jacob and M. Möser. Active control of double-glazed windows. Part 1: Feedforward control. *Applied Acoustics*, 64:163–182, 2003.
[141] A. Jacob and M. Möser. Active control of double-glazed windows. Part 2: Feedback control. *Applied Acoustics*, 64:183–196, 2003.
[142] S. Pietrzko and O. Kaiser. Experiments on active control of air-borne sound transmission through a double wall cavity. In *Proceedings of Active 99*, pages 355–362, 1999.

[143] A. Jacob and M. Moser. Active control of the cavity sound field of double panels with a feedback controller. In *Proceedings of the ICSV7*, 2000.

[144] C. Deffayet and P.A. Nelson. Active control of low-frequency harmonic sound radiated by a finite panel. *Journal of the Acoustical Society of America*, 84:2192–2199, 1988.

[145] C.R. Fuller, C.H. Hansen, and S.D. Snyder. Active control of sound radiation from a vibrating rectangular panel by sound sources and vibration inputs: An experimental comparison. *Journal of Sound and Vibration*, 148(2):355–360, 1991.

[146] T. Bein, T. Doll, and L. Kurtze. Active facades to reduce noise emissions. *Intelligent Glass Solutions*, 3:41–44, 2007.

[147] J. Sagers, T. Leishman, and J. Blotter. Active sound transmission control of a double-panel module using decoupled analog feedback control: Experimental results. *Journal of the Acoustical Society of America*, 128(5):2807–2816, 2010.

[148] T. Bein, J. Bos, L. Kurtze, and T. Doll. Noise insulation applying active elements onto facades. In *Proceedings of the Forum Acusticum*, 2005.

[149] T. Bein, T. Doll, L. Kurtze, and R. Nordmann. Reduction of sound emission in buildings by sound insulation using active facades. In *Proceedings of Inter-Noise 2004*, 2004.

[150] L. Kurtze, T. Doll, T. Bein, and H. Hanselka. Reduced sound emissions of facades using active control techniques. In *Proceedings of the INTER-NOISE 2006*, 2006.

[151] T. Doll, L. Kurtze, C. Langen, and T. Bein. Active structural acoustic control on facades. In *Proceedings of the 13th International Congress on Sound and Vibration*, 2006.

[152] V. Carli, T. Doll, L. Kurtze, and T. Bein. Implementation of an actively controlled system: The active facade. In *Proceedings of the Adaptronic Congress 06*, 2006.

[153] Y. Xun, R. Rajamani, K.A. Stelson, and T. Cui. Active control of sound transmission through windows with carbon nanotube-based transparent actuators. *IEEE Transactions on Control Systems Technology*, 15(4):704–714, 2007.

[154] Y. Xun, R. Rajamani, K.A. Stelson, and T. Cui. Active control of sound transmission through windows with carbon nanotube based transparent actuators and moving noise source identification. In *Proceedings of the American Control Conference*, 2006.

[155] H. Zhu, X. Yu, R. Rajamani, and K.A. Stelson. Active control of glass panels for reduction of sound transmission through windows. *Mechatronics*, 14(7):805–819, 2004.

[156] H. Zhu, R. Rajamani, and K.A. Stelson. Active control of sound reflection, absorption and transmission using thin panels. *Noise and Vibration Worldwide*, 35(5):8–20, 2004.

[157] H. Zhu, R. Rajamani, and K.A. Stelson. Development of thin panels for active control of acoustic reflection, absorption and transmission. In *Proceedings of the American Control Conference*, 2002.

[158] H. Zhu, R. Rajamani, and K.A. Stelson. Active control of acoustic reflection, absorption and transmission using thin panel speakers. *Journal of the Acoustical Society of America*, 113(2):852(19), 2003.

[159] K. Henrioulle. *Distributed actuators and sensors for active noise control.* Ph.D. Thesis, Katholieke Universiteit Leuven, 2001.

[160] P. De Fonseca. *Simulation and optimization of the dynamic behaviour of mechatronic systems.* Ph.D. Thesis, Katholieke Universiteit Leuven, 2000.

[161] P. De Fonseca, W. Dehandschutter, P. Sas, and H. Van Brussel. Implementation of an active noise control system in a double-glazing window. In *Proceedings of the ISMA 21, Noise and Vibration Engineering Conference*, 1996.

[162] C. Bao and J. Pan. Experimental study of different approaches for active control of sound transmission through double walls. *Journal of the Acoustical Society of America*, 102(3):1664–1670, 1997.

[163] J. Pan and C. Bao. Analytical study of different approaches for active control of sound transmission through double walls. *Journal of the Acoustical Society of America*, 103(4):1916–1922, 1998.

[164] P. Gardonio and S.J. Elliott. Active control of structure-borne and air-borne sound transmission through a double panel. *Journal of Aircraft*, 36(6):1023–1032, 1999.

[165] P. De Man, A. Francois, and A. Preumont. Active control of noise transmission through double wall structures: An overview of possible approaches. In *6th National Congress on Theoretical and Applied Mechanics*, 2003.

[166] S. Mohammadi. Semi-passive vibration control using shunted piezoelectric materials. Ph.D. Thesis, University of Lyon, 2008.

[167] D.J. Leo. *Engineering Analysis of Smart Material Systems.* John Wiley and Sons, Inc., 2007.

[168] P.L Reece. *Smart Materials and Structures*. Nova Science Publisher, Inc., 2006.
[169] IEEE Standard on Piezoelectricity: ANSI/IEEE Std 176-1987. IEEE 1988.
[170] R.J. Silcox, S. Lefebvre, V.L. Metcalf, T.B. Beyer, and C.R. Fuller. Evaluation of piezoceramic actuators for control of aircraft interior noise. In *AIAA 14th Aeroacoustics Conference*, 1992.
[171] B. Naticchia and A. Carbonari. Integration of an automated active control system in building glazed facades for improving sound transmission loss. In *ISARC*, 2005.
[172] J.W. Pepi. Fail-safe design of an all BK-7 glass aircraft window. In *Proceeding SPIE, Window and Dome Technologies and Materials IV*, 2286:431–443, 1994.
[173] V. Mallardo, M.H. Aliabadi, A. Brancati, and V. Marant. An accelerated BEM for simulation of noise control in the aircraft cabin. *Aerospace Science and Technology*, 23(1):418–428, 2012.
[174] PPG Aerospace. http://www.ppg.com. 2010. [Accessed on December 1, 2014].
[175] The future by Airbus. http://www.thefuturebyairbus.com. 2011. [Accessed on December 1, 2014].
[176] http://www.feonic.com. [Accessed on March 16, 2014].
[177] E.F. Crawley and J. de Luis. Use of piezoelectric actuators as elements of intelligent structures. *AIAA Journal*, 25(10):1373–1385, 1987.
[178] PI piezo nano positioning. P-010.10P PICA Stack Actuators: Datasheet.
[179] PI piezo nano positioning. PI E-471 High Power Amplifier: Datasheet.
[180] International Standard. *ISO 9614-2 Acoustics — Determination of sound power levels of noise sources using sound intensity*. International Organization for Standardization, 1996.
[181] O. Roth, K.B. Ginn, and S. Gade. Sound power determinations from sound pressure and from sound intensity measurements in a semi-anechoic room. In *Bruel & Kjaer Application Notes*, 2002.
[182] M. d'Ischia, A. Concilio, and A. Sorrentino. Civil-aircraft passenger comfort: Guidelines for analysis and simulation. In *Proceedings of the 4th European Conference on Noise Control, EURONOISE 2001*, 2001.
[183] M. d'Ischia, A. Paonessa, and A. Brindisi. Noise annoyance in civil aircraft cabins: Analysis of an ANN-based evaluation model. In *Proceedings of the 1st Forum on effective solutions for managing occupational noise risks*, 2007.

[184] A. Paonessa, M. d'Ischia, R. Repole, and A. Brindisi. Internal noise and spectral characteristics: Effect of spectral shapes on vibro-acoustic comfort for jet aircraft. In *Proceedings of the XVIII AIDAA National Congress*, 2005.

[185] X. Lin and F.G. Yuan. Diagnostic lamb waves in an integrated piezoelectric sensor/actuator plate: Analytical and experimental studies. *Smart Materials and Structures*, 10:907–913, 2001.

[186] E. Zwicker and H. Fastl. *Psychoacoustics: Facts and Models*. Berlin: Springer Verlag, 1990.

[187] I. Benedetti, M.H. Aliabadi, and A. Milazzo. A fast BEM for the analysis of damaged structures with bounded piezoelectric sensors. *Computer Methods in Applied Mechanics and Engineering*, 199:490–501, 2010.

[188] F. Cotè, P. Masson, N. Mrad, and V. Cotoni. Dynamic and static modeling of piezoelectric composite structures using a thermal analogy with MSC/NASTRAN. *Composite Structures*, 65:471–484, 2004.

Index

1/3 octave bands, 23

accelerometers, 196
acoustic absorption, 24
acoustic disturbance
  stepped-sine, 228
acoustic enclosure, 206
acoustic impedace, 45
acoustic potential energy, 98
acoustic relevance, 182
acoustic resonances, 75
acoustic wave equation, 29
acoustic wavelength, 54
acoustic-structure resonances, 164
acoustically efficient sidewall design, 139
active control of sound, 93
active control of sound transmission, 104
active mounts, 149
active noise cancellation, 93
active noise control, 7
  acoustic metric, 7
  single-channel, 93
  wave interference, 8
active noise control concepts, 12
active noise control performances, 231
active segmented partitions, 147
active sound transmission control, 96

active structural acoustic control, 7
  structural metric, 7
  weak radiator, 8
active structural acoustic control strategy, 158
actuation effort, 120
actuators, 9
  magnetostrictive, 10
  piezoelectric, 10
adaptive controller, 96
adaptive digital filters, 107
adaptive feedforward, 14
adaptive least mean squares (LMS), 96
adiabatic process, 29
aircraft passenger discomfort, 5
aircraft passengers, 2
aircraft window, 139
  acoustically-optimized, 143
aircraft windows, 11
  laminated glass, 144
  plexiglass, 144
  privileged path, 11
airframe, 1
anechoic chamber, 23, 182
anechoic room, 21
  free field, 21
application-oriented strategy, 141
Arnhym, Albert, 9
array of microphones, 98

279

artificial neural network, 241
  multi-layer perceptron, 241

baffled plates, 34
band-limited white noise excitation, 136
barriers, 19
benchmark examples, 83
bi-directional linear sweep, 130
blade passage frequency, 141
blocking force, 155
broadband noise, 94

cavity, 63
cavity absorption, 90
cavity acoustic resonances, 54
cavity control, 98
cavity-dominated modes, 187
coincidence frequency, 22
  dip, 73
collocated feedback control, 93
comfort, 2, 94, 158
  perception, 3
  psychoacoustic, 12
comfort benefits, 239
comfort bubble, 158
comfort index, 241
comfort predictions, 191
commercial off-the-shelf, 150
control, 5
  active, 7
  feedback, 8
  feedforward, 8
  passive, 7
  semi-active, 7
control authority, 221
control effort, 218
control microphone, 208
control objective, 98
control strategies, 12
control system
  brain, 101
controller, 93
convergence coefficient, 128

counteracting acoustic field, 93
cruise conditions, 207
curve fitting methods, 25

damping, 54
data acquisition, 215
destructive interferences, 147
diffues
  incident waves, 83
diffuse sound field, 80
digital filters, 14, 96
digital signal processing, 7, 94
Dirac delta function, 32
directivity, 20
Dirichlet, 30
discomfort, 221
dodecahedron speaker, 192
double pane aircraft-type windows, 139
double-panel
  finite, 76
  infinite, 55
DSP control board, 190
DSP implementation, 206
dynamics of structures, 17

efficient radiators, 44
eigenfrequency, 218
eigenvalue problem, 26
  characteristic equation, 26
  generalised coordinates, 27
  modal expansion, 27
  undamped vibration, 26
elastic suspensions, 58
electro-mechanical coupling, 147
electronic controller, 105
energy dissipation, 90
engine run-up, 131
  adaptive algorithm, 131
engines, 1
equations of motion, 26, 49
error signal, 93, 211
Eulerian FE/FE model, 50

INDEX

experimental modal analysis, 25
  modal parameters, 25
  natural frequencies, 25
  normal-mode, 25
experimental model-based approach, 115
experimental system identification, 114
experimentally-oriented research, 156

failsafe design, 157
far field, 36
feedback active noise control, 93
feedback control
  loop gain, 103
  PID-control, 104
  sensitivity function, 103
  stability requirements, 103
feedforward active noise control, 93
feedforward control
  pure open-loop, 101
feedforward simulations, 119
field
  incident field, 4
  near, 4
  radiated, 4
filter coefficients, 122
filtered-reference LMS algorithm, 110
filters
  antialiasing, 226
  reconstruction filters, 226
flanking transmission, 198
fluid gap, 53
fluid-matrix system matrices, 50
fluid-structure interaction, 13
  coupled, 13
  uncoupled, 13
force appropriation method, 25
  normal mode testing, 25
free field radiation, 55
frequency response function, 25
frequency spectrum, 224

Fuselage panels, 4
  double-walled, 11
  trim, 10
fuselage shells, 189

Galerkin formulation, 46
Green function, 32
Green's first identity, 46
Green's second identity, 67

harmonic excitation force, 26
hearing sensations, 239
Helmholtz, Hermann, 29
high convergence rate, 222
high modal density, 99
human perception, 190
human sound perception, 239
human-centred design, 159

impact modal test, 174
  accelerometers, 175
  load cell, 175
impulse response, 107
incidence transmission
  normal, 73
  oblique incidence transmission, 73
incident panel, 63
inhomogeneous Helmholtz equation, 31
integrated circuits, 94
intensity levels, 233
  active, 233
  reactive, 233
intensity probe, 231
interior trim panels, 158
iterative optimization, 102

Kirchoff–Helmholtz, 51
Kirchoff–Helmholtz equation, 34

laboratory set-up, 194
LMMS, 113
LMS algorithm, 109
low-frequency sound insulation, 140
Lueg, Paul, 9

mass law, 90
mass-air-mass, 13
mass-air-mass resonance, 54
MELMS, 113
microphones, 147, 195
middle panel, 63
modal coupling theory, 13, 55
modal restructuring, 95, 146
modal suppression, 95, 147
monopole, 31
multi-channel adaptive algorithms, 111
multi-channel algorithms, 14
multi-channel ASAC controller, 136
multi-functional materials, 151
multi-input-multi-output (MIMO), 94
multi-output control, 220
multi-panel partitions, 13, 85
multi-wall structures
　plate-like multi-wall structures, 55
multiple partitions, 53
mutual-radiation resistances, 98

natural frequency analysis, 199
N degree of freedom system, 57
neighbouring, 236, 237
Neumann, 30
no reflections, 30
noise, 1
　airborne, 3
　interior, 1
　radiated, 5
　sources, 4
　structure-borne, 3
noise and vibration, 2
noise reduction, 94
normal incidence mass law, 56
numerical methods, 18
　BEM, 18
　FE-BE, 18
　FE-FE, 18
　FEM, 18
　SEA, 18
numerical tool, 79

observation point, 109
octave, 23
off-line identified plant response, 211
off-resonance active control, 147
on-line identification, 95
on-resonance case, 218
out of phase, 88
overdetermined case, 100

panel partitions
　double, 39
　triple, 39
passenger level, 239
passenger perception, 239
passenger-focused approach, 141
peripheral paths, 206
phase separation technique, 25
　circle fitting method, 28
　complex exponential method, 28
　peak amplitude method, 28
　polyreference complex exponential method, 28
piezo actuators, 122
piezoceramic stack actuators, 148
piezoelectric patches, 7, 147
piezoelectricity, 151
　direct piezoelectric effect, 151
　inverse piezoelectric effect, 151
　piezoceramics, 153
　polyvinylidene fluoride polymer, 153
　stack actuators, 153
piston-like coupled mode, 169
planar vibrating plate, 35
plane wave, 80
plane wave propagation, 70
polyvinyl fluoride (PVDF), 95
primary disturbance, 93
primary noise source, 122
primary plant, 93
psychoacoustics, 239
pure active vibration control, 97

quantitative metrics, 19

INDEX

radiating panel, 63, 141
radiation modes, 41
radiation reactance, 40
random noise, 123
rate of flow, 231
Rayleigh integral, 13, 35, 80
receiving room, 20
reference signal, 93, 208
reference test suite, 179
reflected wave, 53
resonance peak, 205
reverberant spaces, 80
reverberation chamber, 23
    diffuse, 23
reverberation time, 24
rigid boundary, 45
rigid-walled cavities, 164
Robin, 30
robustness, 95, 222

scanning error-LMS, 114
secondary path, 94, 109
secondary source, 94
    perfectly coherent, 123
self-radiation resistances, 98
sensors, 9
    microphones, 10
shape function, 47
shell elements, 83
short time Fourier transform, 228
simple test cases, 92
simply supported, 36
simply supported panel, 84
single output real-time controller, 220
single-input-single-output (SISO), 94
smart panels, 12, 119, 141
    cavity control, 146
    panel control, 146
    room control, 146
smart structures, 2
    market catalysts, 2
Sommerfeld, 30
sound calibrator, 211
    piston-phone, 211
sound energy, 196

sound insertion loss, 21
sound intensimetry, 98, 190
    discrete point method, 232
    scanning method, 232
sound intensity, 23
sound intensity level, 231
sound intensity map, 235
sound intensity probe, 196
sound path, 186
sound power, 37
    elemental radiators, 37
sound pressure level, 12, 19, 231
sound radiation, 17
    cross-modal, 38
    efficiency, 42
    free space sound radiation, 33
    self-modal, 38
sound transmission, 69
sound transmission loss, 22, 53
sound transmission simulation
    procedure, 81
sound transmission suite, 189
soundproofing, 144
source room, 20
source strength, 31
spatial localization, 233
spectogram, 135
spectrum analyzer, 185
spring-loads, 165
spurious sources, 123
standing waves, 88
state-space models, 117
steady-state acoustic pressure, 48
steepest-descent algorithm, 105
stochastic disturbances, 108
structural acoustic coupling
    weak, 18
structural dynamics, 26
structural modal excitability, 178
    grouped piezo stacks, 178
    single actuators, 178
structural modes, 41
structural path, 185
structural wave speed, 72
structural-acoustic coupling, 17

subjective, 240
subjective interior comfort, 241
subjective parameters, 239
subjective quality, 239
subjective response, 239
   human, 240

tachometer, 141
test-bed, 157
thermal active power, 172
third-octave bands, 127
threshold, 225
time invariant, 228
tracking performance, 222
transfer functions, 95
transmission
   transmission paths, 4
transmitted wave, 53
transparent actuators, 148
triple pane windows, 139
triple partitions, 69
turboprop, 5
turboprop aircraft cabin, 119
   three-bladed propeller, 119

vibrating boundary, 45
vibration absorbers, 11

vibro-acoustic comfort, 239
vibro-acoustic environment, 241
vibro-acoustic problem, 62
vibroacoustic comfort
   objective, 240
virtual passenger model, 13, 240
virtual passenger response model, 241
viscoelastic damping, 143
volume velocity, 34

wavenumber, 70
waves
   uncorrelated, 80
weak coupling, 80
well conditioned, 213
white noise, 117
Wiener filter, 108

zone of quiet, 93
Zwicker's parameters, 240
   fluctuation, 240
   loudness, 240
   roughness, 240
   sharpness, 240
   tonality, 240

Printed in the United States
By Bookmasters